CULTURE AND HUMAN-ROBOT INTERACTION IN MILITARIZED SPACES

Emerging Technologies, Ethics and International Affairs

Series editors:

Steven J. Barela, University of Geneva, Switzerland
Jai C. Galliott, The University of New South Wales, Australia
Avery Plaw, University of Massachusetts, USA
Katina Michael, University of Wollongong, Australia

This series examines the crucial ethical, legal and public policy questions arising from or exacerbated by the design, development and eventual adoption of new technologies across all related fields, from education and engineering to medicine and military affairs.

The books revolve around two key themes:

- Moral issues in research, engineering and design.
- Ethical, legal and political/policy issues in the use and regulation of Technology.

This series encourages submission of cutting-edge research monographs and edited collections with a particular focus on forward-looking ideas concerning innovative or as yet undeveloped technologies. Whilst there is an expectation that authors will be well grounded in philosophy, law or political science, consideration will be given to future-oriented works that cross these disciplinary boundaries. The interdisciplinary nature of the series editorial team offers the best possible examination of works that address the 'ethical, legal and social' implications of emerging technologies.

Culture and Human-Robot Interaction in Militarized Spaces

A War Story

JULIE CARPENTER

Ethics + Emerging Sciences Group,
California Polytechnic State University, USA

Routledge
Taylor & Francis Group

LONDON AND NEW YORK

First published in paperback 2024

First published 2016 by Ashgate Publishing

Published 2016 by Routledge
4 Park Square, Milton Park, Abingdon, Oxon OX14 4RN

and by Routledge
605 Third Avenue, New York, NY 10158

Routledge is an imprint of the Taylor & Francis Group, an informa business

Publisher's Note
The publisher has gone to great lengths to ensure the quality of this reprint but points out that some imperfections in the original copies may be apparent.

British Library Cataloguing in Publication Data
A catalogue record for this book is available from the British Library

The Library of Congress Cataloging-in-Publication Data has been applied for

ISBN: 978-1-4724-4311-3 (hbk)
ISBN: 978-1-03-292845-6 (pbk)
ISBN: 978-1-315-56269-8 (ebk)

DOI: 10.4324/9781315562698

I dedicate this work to my great-aunt Alma P. Mayer and my friend Dr. Clifford I. Nass, two people who taught me the value of listening.

Patrol work was desperately dangerous in the last war. But a flying robot, equipped with motion picture and sound recording machines, could dart and hover over the enemy with no danger to human life and bring back vastly more accurate observations. When a human soldier gets a bullet in his heart, or in his liver, or has himself partly blown to pieces, that is the end of that soldier. Not so with the robot. A new heart can be put in him as easily, almost, as changing a tire.

— *San Antonio Light* (1935)

Contents

List of Figures

About the Author

Julie Carpenter worked in user experience testing, human-computer interaction, and the research and development of Web-based applications before pursuing her interests in human-robot interaction. Her curiosity about how people interact with technology inspired her to earn an M.S. in Technical Communication from Rensselaer Polytechnic Institute, a second M.S. from the Technical Communication program at the University of Washington, and a Ph.D. in Learning Sciences from the University of Washington. Her principal research interest is human-robot interaction, specifically human emotions and attachment to robots, especially in field applications (for example, space exploration, defense, and humanitarian relief efforts). She has published a number of peer-reviewed articles concerning topics such as the rhetoric of humanoid robot design, (US) Explosive Ordnance Disposal personnel interactions with field robots, and human-centered research in robotics and other new technologies. Currently, she is a Research Fellow in California Polytechnic State University's Ethics + Emerging Sciences Group.

In her spare time, Dr. Carpenter enjoys travel, gardening, baking, reading, and her many and assorted pets. She lives in the Pacific Northwest, and considers this beautiful part of the country her home.

Foreword

We are at the dawn of the age of the robot.

Of course, we have been successfully utilizing robots for decades now. Robotic spacecraft have mapped the Earth, helped forecast storms, probed other planets, and successfully landed on an asteroid. This was the easy part in robotic development, because these robots did not have to interact with *us*. They were hundreds or thousands of miles away; we could not see them. They were abstract. Now they are starting to work with us, down here on the real world.

These robots are not very smart or very capable, but neither was the Apple II microcomputer when it was introduced in 1977, with a 1-megahertz processor, 4 kilobytes of RAM, an audiocassette interface, and a retail price of $1,298. Today, the $849 iPhone 6S has a 1.8-gigahertz processor, 128 gigabytes of memory, and our biggest complaint is that it may bend in the back pocket of our skinny jeans. Today's robots are at the Apple II level—maybe at the level of the Apple II+.

Ground robots will develop at the same rate—probably faster. The intersection of artificial intelligence, compact and efficient batteries, and motion technology is coming more quickly than we may be ready for. Luminaries such as Elon Musk and Dr. Stephen Hawking have recently declared artificial intelligence as the technology that may end the human race. They may be right, if our study of human interaction with evolving technology lags the exponential advances we are likely to see. But history shows if we pay attention, we can still get it right.

Julie Carpenter, with this book, is helping us get it right.

Militaries typically have been slow to adapt evolving technologies into commonly accepted tactics, frequently at their peril. Lieutenant Colonel George Custer failed to take available Gatling machine guns to his encampment at Little Big Horn, which, if used, may have saved his battalion from extinction at the combined hands of the Lakota, Northern Cheyenne, and Arapaho warriors. Although the British invented the tank, much of the world's cavalries remained on horseback until the Germans successfully integrated battle tanks into their deadly successful blitzkrieg technique that brought much of Europe to its knees in the late 1930s and early 1940s. Perhaps the most potentially dangerous example of failure to adapt a new technology was the successful development and fielding of nuclear weapons in 1945. For years and years, the Pentagon viewed nuclear weapons as just another big bomb that could be integrated in our war-fighting strategy, despite the likelihood of a humanity-altering result. My father still recalls his F-105 fighter jet being equipped with air-to-air nuclear missiles in the late 1950s; the idea was to explode a nuclear warhead in the path of a Soviet bomber formation. There were several drawbacks to this strategy, not the least of which was the electromagnetic pulse (EMP) that would have also destroyed my father's jet. It was not until the early 1970s that the nuclear powers of the world began to view nuclear weapons as something to have, not something to use, and they are now viewed primarily as a deterrent. As with the lesson of nuclear weapons, it is important the military develop policy for employment of robotics, particularly because of the rapid increases in robotic technology. Robots will not only change *how* we fight wars, but they will also change *who* does the fighting.

As an Army officer, my job is to provide national security for the citizens of the United States. When war, a violent action similar to war, or operations other than war are decided upon by Congress and the president, my job is then to figure out, in association with my brothers-in-arms from the Army and our sister services, how to successfully bring hostilities to a conclusion while simultaneously accomplishing the national security objectives set forth by our elected and appointed civilian leaders. Implied in that mission is that we accomplish these objectives with minimum loss of lives to United States citizens and all non-combatants. Here, while robots cannot yet replace human soldiers, they can begin to take on the most dangerous and lethal tasks of war.

Dr. Carpenter spent an extensive amount of time interviewing members of the Explosive Ordnance Disposal (EOD) community, today's most prolific military users of ground robots, to determine how users interact with their robots in combat and non-combat environments, and to make predictions about how these interactions might change going forward as ground robots become more anthropomorphic in structure, so they can be operated on more varied terrain, go everywhere the soldiers go, and use the same tools as they do.

By studying the social science side of this issue now, the academic and scientific community will be able to keep up with the human factors associated with the exponential technological advancements in artificial intelligence and robotics. Dr. Carpenter's personal observations and research findings have implications for the future integration of human and robot soldiers, as the trend towards a larger role for autonomous and semiautonomous machines to perform the U.S. military's dull, dirty, and dangerous jobs is likely irreversible. Enjoy the book!

Lieutenant Colonel Michael Kolb, Ph.D.
National Geospatial-Intelligence Agency
Washington, D.C.

Preface

The future of warfare is, in many ways, about forging the new culturally embedded particulars of successfully working with robots. Building, designing, and using robots for defense. Revised practices that include robots as heavily integrated parts of troop training, as well as techniques, tactics, and procedures. Negotiating and adhering to ever-changing legal and ethical standards held in place by local, federal, and international entities. Preparing for and responding to proposed policies by non-governmental organizations that gain wide support and bring intense scrutiny to operations that use robots, especially in combat. The economics of wide-scale purchases for a technology that is constantly evolving. Individually and collectively, constructing new norms, customs, and expectations of how to interact with robots, on the battlefield and off. Integrating medicine and aid with robotics, and developing new models of psychological trauma that include the close integration of robots in personal experiences. The variables of humans, robots, and war are becoming entwined into an endless twist of possible paths.

The very idea of beginning to understand the effects of new tools in defense work is a daunting one, and in this book, by and large, I choose to refer to robots as *tools*. When I first began exploring the uses of robots in the military, I was reluctant to categorize robots as anything but *robots*, in my mind or out loud. To me, robots were too robust—and their changing roles in our world too new—to fit into any category that already existed. I was also consciously attempting to remove my own preconceived ideas of robot roles in defense before I began talking to the people that work with robots every day. Research into the topic has demonstrated there are many ways people tend to categorize robots depending on their perceived or presented primary role, such as assistant, tool, teammate, companion, or servant. However, when doing the research for this book I was immersed in conversation with military personnel who used robots every day, and *tool* was their primary word for referencing robots; so it became my own, at least for now, and in this context.

All research contains bias and influence from the points of view of those who choose the topic to study and the questions to ask, although we attempt to remove ourselves from the final reported findings as if they spontaneously erupted from a laboratory insulated from the outside world and its influences. Yet the methods of inquiry, modes of data analysis, and how findings are presented are all part of a series of choices we make as investigators, for a variety of reasons both practical and personal. The very language and style we use in reporting our findings—in third-person, straightforward, and without literary flourish—is supposed to further frame our work as clean and untouched by any outside influences.

The peer-review process is one way researchers negotiate the relative truth of their findings. As investigators, we meticulously note our methods throughout the course of research, cross-check findings with others working with the same data, and apply a framework of rigorous, peer-accepted protocol to every step of the process. Knowing we inherently have our particular ways of looking at questions and answers, we also attempt to be transparent in our procedures so others can replicate findings to prove or disprove them, and build on the existing work. Across academia and science, these types of conventions generally serve their purposes and provide a scaffold of theory, methods, analysis, and reporting that presents what we see or hear in the truest way possible, or at least, the truest way we can.

As individuals and as part of a larger social system, researchers determine scientific truths through constant negotiation via a system of checks and balances from within our own experiences and those of our community of colleagues. It is one reason scientists are trained to regard one another's work with such critical eyes. What if she had chosen another method of inquiry? What if the next study used a different criteria set for study participants? What if he had asked different research questions at the outset? What if that study began an investigation without overarching research questions at all, but rather found patterns in the data in order to learn what questions might be useful in the future? Which of these things should or can be addressed in the next iteration of study on this topic? Yet, ultimately, it will always be the researchers' version of truth. Still, research biases of the sort we try so rigorously to avoid do not necessarily make the work less valid or in vain; it does mean as investigators that we are responsible for looking outside our own work continually, comparing versions of truth until we have synthesized our own evolving systems of beliefs. In other words, I think we have to constantly question ourselves, even as we ask research questions of those around us.

All of the people I spoke to for this book that served in the military offered me their truth when I asked for it. It is always difficult to talk about war. Even subjects that might seem to lack some of the immediacy of war's actions or the lingering effects of its transformation, such as the everydayness of it as work for some people, can elicit strong responses. Therefore, I think it is only fair that after I have asked so many people to share remarkable experiences with me that I explain a bit of my own story. Maybe turnabout is fair play. I also happen to believe what a person chooses to share reveals a great deal about their way of seeing things, at least in that moment or space or time.

My mother is first-generation German-American, and the generation before hers are Holocaust survivors and victims. Many of my father's family, also Jewish, escaped Russian pogroms, wandering around Eastern Europe searching for a safe place to land until migrating to the United States in the late 1800s. With this history, I grew up living with people affected by war, and it filled me with questions. As a child, I listened closely to the family history from my grandparents, great-aunts, and uncles, noting the names of who survived, when they immigrated to America successfully, and how they built new lives. I also listened to the family stories about the others, the family that did not survive. Some were killed in concentration camps, or died in forced migration, or did not survive brutal ghettos. Some of these family members looked down on me from black-and-white and aging sepia-toned photos hung on the walls of my grandparents' house. I spent years looking at their faces, trying to imagine what they were like, what they went through in everyday life. However, these ghosts in my life are not as morbid an influence as it might at first read. In fact, I think of all of these people as close family, passing on their knowledge and energy to me via my family's vivid oral histories.

Even so, I also know the generation that included my grandparents and great-aunts and great-uncles tended to omit many of the details when retelling the sadder stories to me because they could not bear reliving the experiences without swells of overwhelming emotions. I am sure they also wanted to protect me, as their grandchild or great-niece, from some of the brutality of their experiences.

Sometimes I would learn the more mournful pieces from another relative, or from my own digging through family documents from surviving artifacts, asking my parents to fill in gaps of information, and then putting all these puzzle pieces together. When more than one person was narrating a past event, sometimes other relatives would chime in with details, opinions, and corrections. Moreover, when each tale intertwined with the other, my family created a rich living history for me, albeit often in chapters I mentally had to reassemble so I could give the complex and tangled stories a sense of order, a time line that I could follow in my mind. Through this storytelling, collections of photos, and family documents, I feel lucky to have a sense of personal history and of knowing some family members without ever having met them. I know I come from cattle traders, farmers, bakers, butchers, linen cleaners, carpenters, academics, rabbis, mothers, fathers, wives, husbands, daughters, sons, survivors, who each have a story.

Of course, these circumstances raised some very serious questions for me, even as a young child. Why would someone want to hurt my family? Are we bad people? Are the people that hurt us bad? What separated *us* from *them* in such a divide that we were systematically sought out as targets to be eliminated? What made some human beings decide other human beings were less than human, or could be treated in inhumane ways? What defines a human and humanness? These are enormous concepts to begin pondering as a child, and I know I am—unfortunately—hardly the only person from a family ravaged by war who had similar thoughts run through her head.

In my mind, the good I can take from my personal background is that it was formative for my deep interest in people and their stories. Through this experience, it makes sense I was drawn to personal narratives, emotions, and their impact on people's everyday lives. Listening to someone reveal pieces of his or her life is a privilege, and one I do not take lightly.

From Explosive Ordnance Disposal (EOD) personnel, I learned that the nature of their work is unique within military specializations in many ways, as a career path and at a very personal level. Even their initial training period includes a component that all members of the Armed Forces attend together in Eglin Air

Force Base (AFB), so they begin with a shared touchstone of experience from their earliest introduction to EOD work.

As the nature of EOD training and work evolves within the military, and because of a surge in Improvised Explosive Device (IED) encounters, modifications are being made to many aspects of EOD teamwork. These include team size, the age of people accepted into the program, and an increased reliance on technology such as robots. One of the most critical standard tools EOD personnel use are the semiautonomous teleoperated robots that assist in Render Safe Procedures (RSP), helping to disable or mitigate the threat of explosives. Accordingly, if problems with the human-robot interactions are overlooked, there is a continued danger to human lives and mission outcomes from the unidentified issues in these interactions.

Up until the study described in this book, there has been no inductive research approach to investigating the dynamics of the EOD personnel's interactions and experiences with robot models used every day or the associated emotional aspects of these interactions. These interactions include how emotion in human-robot interaction affects operator decision-making and, therefore, mission outcomes. I will not downplay my advocacy for interdisciplinary robotics research, because I strongly believe it to be necessary. EOD work is only one example of how interacting with robots every day is life changing for the people in the equation.

It is widely acknowledged that in human-human relationships, the process of attachment and its related concepts of bonding, cohesion, and trust can influence our actions and behaviors. I believe discovering which, if any, of these human factors play into the human-robot dynamic will lead to insights about ways of leveraging the robot design elements or contexts of use that trigger positive and negative operator reactions. Additionally, in order for human-robot teams to be effective, I felt research was needed into the whole system that the individual team members are a part of, and into how these factors ultimately shape and influence the interactions at the micro levels. In order for the desired outcomes—successful missions and safe personnel and civilians—EOD human-robot interactions need to be as fluid as possible, so both humans and robots can overcome obstacles efficiently.

When considering my research approach, I believed a holistic picture of the social systems interactions would provide a useful base knowledge that could stimulate the identification of factors that affect the human-robot experiences. With this knowledge—combined with pieces revealed in others' research—we can improve human-robot training, advance robot design, and develop effective support of mission interactions between humans and robots. Accordingly, to develop a basic understanding on which any coherent discussion of robot design and use within close teams such as EOD can be based, my own initial study delved into the human user experience of EOD human-robot interactions. To achieve these goals, the study examined the context, expectations, attitudes, and emotions that are part of these human-robot relationships.

I prefer an ethnographic approach for initial forays into new territory, and describe phenomena as a form of explanation, a beginning point for gaining clarity and discovering what questions we might ask next. The nuances of human behavior toward machines are not easily categorized or predicted, unfortunately not leaving a neat map or series of if/then statements for us to follow. My friend and colleague Dr. Emma Rose is fond of referring to this aspect of human-centered design research as "squishy" because of its abstract beginnings. Generally, it is has always been the squishy things in human nature I find most interesting in relation to our interactions with technology. In a formal study about subjective experiences, like this one, we rigorously investigate people's thoughts and ideas, internal processes that do not produce easily observable data but exist, nonetheless. Therefore, our first goals are to seek out these thoughts and personal experiences, the squishy things, by asking questions and recording narratives. Then, we seek patterns in these accounts in the hope they will lead to the more concrete or actionable steps.

With rich descriptions in place of what has happened and what is happening now in relation to phenomenon, we are armed with a better understanding of what questions to ask next, and it is a foundation we can to continue to build upon. Based on this premise, this book is divided into three parts: *Narratives* (Part I), *Metaphors* (Part II), and *Patterns* (Part III). The chapters that follow are a framework for a discussion of the enormity of the pieces in play with a focus on the human perspectives and experiences.

My own emotional influences have converged and guided my actions toward a deep interest in things related to human communication, technology, culture, and history. Therefore, I want to acknowledge some of the people who have influenced and supported me along the path I have taken so far.

I am grateful for the wisdom of my teachers, friends, and colleagues at the University of Washington: John D. Bransford, Leslie R. Herrenkohl, and Stephen T. Kerr. Alexandra L. Bartell and Joan M. Davis have been dear friends and tireless cheerleaders for my interests and efforts. Clifford I. Nass of Stanford University showed me many kindnesses over the years, and I will always remember his generosity of spirit and mentorship.

Jai C. Galliott reached out to me as a series editor at Ashgate, and for that first step toward this book, I am grateful. Brenda Sharp and Philip Stirups cheerfully and patiently answered all my questions about the publishing process, and I very much appreciate their guidance. Andrew Tedlow good-naturedly supported me during the long process I went through writing this book. Patrick Lin listened to my fears while I wrote, and I will thank him here even though (or because) he insists Peter Gabriel lyrics are the answer to everything. While preparing this book I connected with Jeff Vintar, who instantly felt like a natural part of my life in the constellation of our science fiction and research and writing and real worlds. Thank you, Vintar the Amazing, for your magic.

Of course, I am thankful to my family for their traditions of storytelling, and my mother for her encouragement. In particular, working on this book made me think of my great-aunt Alma, who, in her gentle way, shushed other adults just so she could listen to my brother and me play when we were children. Aunt Alma always made me feel like she heard more than what I said, in the best possible way of understanding.

During the data-collection phase and even after my dissertation's publication, many people sought me out of their own accord to share their own relevant experiences, whether they participated formally in the study or not. To them, I say I am so very thankful for their time and trust. Lastly, I feel I cannot emphasize this enough: none of my work would be possible without the people who shared their stories with me for the study, and I wholeheartedly thank them for putting their thoughts and feelings into words.

The complex social system of human-robot interaction in militarized spaces is vast, and this book attempts to connect different aspects of human relationships to technology within the unique spaces and contexts of defense work. These chapters and sections are not meant to be an exhaustive examination of possibilities, but rather a means for exploring different parts of the systems that have come in to play. Read this book as an attempt to raise new questions, rather than as a way to settle upon one path.

Julie Carpenter,
Portland, Oregon
January 2016.

Glossary of Acronyms

AFB	Air Force Base
AI	Artificial Intelligence
C-IED	Counter-Improvised Explosive Device
Capt	Captain (Air Force, Marine Corps)
CPT	Captain (Army)
DARPA	Defense Advanced Research Projects Agency
DoD	Department of Defense
EOD	Explosive Ordnance Disposal
FOB	Forward Operating Base
HBIED	House Born Improvised Explosive Device
IED	Improvised Explosive Device
IND	Improvised Nuclear Devices
JIEDDO	Joint Improvised Explosive Device Defeat Organization
K-9	Canine, or Military Working Dog
LARS	Lethal autonomous robots
LAWS	Lethal Autonomous Weapons
Maj	Major (Air Force, Marine Corps)
MAJ	Major (Army)
MGySgt	Master Gunnery Sergeant (Marine Corps)
MOS	Military Occupational Specialty
MSG	Master Sergeant
MWD	Military Working Dog
NCO	Noncommissioned Officer
PO1	Petty Officer First Class
RPA	Unmanned or Unpiloted Aerial Vehicle
RPV	Remotely Piloted Vehicle
RSP	Render Safe Procedures
SCPO	Senior Chief Petty Officer
SOF	Special Operations Forces
SSG	Staff Sergeant (Army, Marine Corps)
SSGT	Staff Sergeant (Air Force)
TSgt	Technical Sergeant
TTP	Tactics, Techniques, and Procedures
UGV	Unmanned Ground Vehicle
USAF	United States Air Force
USMC	United States Marine Corps

USN	United States Navy
UXO	Unexploded Ordnance
VBIED	Vehicle-borne Improvised Explosive Device

PART I
Narratives

The world has been looking forward to robots for a long time.

Judging by the stories we have told throughout history—and continue to tell—every culture has imagined how we might interact with artificial life in different forms. Fictional stories can persuade and paint vivid pictures and propel ideas in a strong way, especially as an introduction to a set of complex concepts, such as how we might interact with robots. Whether it is tales of fantasy creatures, aliens, vampires, ghosts, or robots, people around the world enjoy speculating how and why and when relationships between humans and these other things might happen, with all the associated pitfalls, adventures, expectations, and questions.

For good or bad, there is no way to move through life without the influence of fellow humans. How people come to know and interpret real objects and events is the result of ongoing ways of understanding surroundings based on individual experiences. Yet individual experiences include interacting with others. The way people learn to assign significance and meaning is not developed in each person alone, but in coordination with other people. In adulthood, some of the social groups that influence people every day include families, friends, workplace or professional organizations, and coworkers, all in varying degrees. The way individuals absorb or reject these influences is sometimes a conscious effort, and requires purposeful learning and adaptive thought processes to include new understandings. Perhaps more often, people come to accept certain everyday conventions without devoting too much analytical thought to how they arrived at a particular set of beliefs, values, and even expectations about others.

The ways people have historically talked about subjects like war and the tools of war are important to consider because stories are powerful. A story can be something told as *real*, from one person's perspective, a retelling of events. Other stories come from outside a first-person experience, and may be from another person's viewpoint of the same event. Of course, some stories that resonate are meant to be regarded as total fiction, and share events in a way that is entertaining. A story is really a way to present the storyteller's agenda or way of thinking, and carries messages from teller to listener. The process of changing one mind or many can begin with a moving story, a narrative that captures imagination and taps into human emotions. History, of course, is its own way of collective storytelling, sharing a version of events that communicates about the past. By systematically and critically reviewing history, stories, accounts, anecdotes, and narratives, it is possible to gain great insights about people's expectations, goals, actions, and behaviors in certain contexts and circumstances.

Therefore *Narratives* begins with a framework of the significance of cultural studies in defense and robotics. Then, in order to understand where defense robotics are now, there is an overview of the U.S. military research's recent historical relationship with experimental robotic technology. One way to trace the advancement in this arena of warfare is by examining the inventions and uses of militarized robotic systems. Tools, and the creation and use of tools like robots, can significantly change feelings about war at a global level, fueling an ongoing discussion about policy, law, training, civilian casualties, and the changing roles of human personnel. In the case of robots and warfare, these global discussions often center on the difficult questions of how and when robots will be used. Although the exact questions will change as our use of robots becomes more of an everyday reality and is integrated with human teams at every level of military work, critical thinking will (and should) be ongoing due to the enormous changes and rippling impact robots bring to the acts of war.

Chapter 1
Learning by Experience

Culture War

Culture is one of the defining characteristics surrounding what it means to be human. A complicated set of ideas and actions, culture is the range of phenomena circulated through social learning. In a general sense, culture is the ability to categorize and portray experiences with symbols and to act imaginatively. Although these cultural processes and practices are not unique to humans as some other species have demonstrated aspects of social learning, humans are in many ways reliant on culture to frame values, beliefs, and expectations about the world, influencing individual and collective behaviors.

Culture manifests as complex relationships of practices stemming from collective knowledge transmitted through specific group social interactions, and the symbolic constructions that give those activities meaning and significance. Traditions, law, customs, social standards of conduct, and religious beliefs are all examples of cultural components. Physical expressions of culture, such as technology or art, may be considered part of material culture, artifacts loaded with clues about the people and groups that design, make, and use them. The principles of *social organizations* are perhaps less tangible, but still a key part of the idea of culture. Philosophy, literature, mythology, science, and art are all pieces of the *cultural heritage* of a society or group.

The human abilities to classify, categorize, encode, and decode experiences symbolically, and be able to communicate these encoded experiences socially, are parts of culture. In sociology, culture is further defined as ways that thinking, behaving, and material objects—sometimes referred to as artifacts—combine to mold the ways in which people live their lives. Using these concepts, these artifacts are considered meaningful objects for study of cultures through their choices of design, efficacy, materials used, engineering, production, sharing, and uses.

Because of the speed with which robots are becoming incorporated into everyday lives, there is an increasing need to explore the notion of *robotness* as well as *what it means to be human* in this discovery phase of cultural exploration in human-robot interactions. Studying culture and moving forward in an understanding of our humanity and humanness also helps define *robotness*, or a similar descriptor that can apply to what it means to be a robot beyond a machine. The development of a new technology is a factor that influences complex societal changes by shifting social dynamics and creating new cultural models. These social changes accompany ideological alterations and other types of cultural transformation.

For example, cultural movements involve new practices that produce a shift in relations between groups, influencing both social and economic structures.

Contact between societies also affects culture, by engenerating or inhibiting changes in cultural practices. For example, competition over resources and war influences technological development and large and small social interactions. Ideas are transferred and shared between cultures, and the artifacts of culture can be shared, although sometimes the meanings attached to those objects change as they become adopted by and integrated into a new group, or are even reinvented altogether.

Modern U.S. Military Humanoid Robot Development

Post-World War II, global competition between countries to discover and introduce the most effective technologies in the fastest development routes led to inventions such as spy satellites, human-portable missiles, and other devices still used in modern combat operations. Robots were included in the research-funded development as a futuristic—yet entirely feasible—technology with many potential uses in warfare. However, it was not until advances in the development and production of different technical systems used by robots that research could really propel forward, building on that progress.

Although the word *robot* has many interpretations, in this book, the term *robot* refers to an embodied mechanical system that acts in an environment or social space, and can interact with people in the physical world. Robots with humanlike physical or behavioral characteristics are commonly referred to as *humanoid*, *humanlike*, or *android*. Those terms may evoke images of highly anthropomorphic robots when actually there are a variety of robots with humanlike characteristics that can be categorized with those terms.

From 1983–1988, the U.S. Space and Naval Warfare Systems (SSC) developed Greenman, a humanoid robot used for remote presence demonstration (Chatfield, 1995). Greenman was operated via teleoperation, with clawlike hands. Even lacking force or tactile feedback, users with little training could perform manipulative tasks using Greenman. Nevertheless, the robot's design also demonstrated the potential usefulness of dexterous hands, force feedback, underwater capabilities, and a high-resolution vision system necessary for aiding divers (Chatfield, 1995).

One of the more widely known early U.S. military anthropomorphic robot research projects produced Manny (Yost, 1989), a humanlike robot with a working artificial respiratory system, but no autonomy or intelligence. Developed to test protective clothing in simulated conditions that are hazardous to humans, Manny was developed for the U.S. Army's Dugway Proving Ground (Fisher, 1988) by Idaho National Laboratory. The robot was 5'11" and 187 pounds, using a clear plastic layered over aluminum forms in a humanlike shape with a skin exoskeleton that registered 98.6 degrees Fahrenheit. Operated by computer, Manny's 38 joints allowed him to walk at three miles per hour, as well as sit, crouch, wave, and crawl (Associated Press, 1989).

Figure 1.1   Manny

Source: Pacific Northwest National Laboratory/BATTELLE, n.d.

In a six-week exhibit at the Chicago Museum of Science and Industry, Manny went through a repetitive series of actions in a 10-minute routine that included "drop-kicking a ball, walking, crouching, duckwalking, and explaining his actions in a clipped robotic voice before folding into the pose of Auguste Rodin's 'The Thinker'" (McEntee, 1989). In the same newspaper article, a program manager explains Manny's realistic humanlike design as functionally "necessary to ensure that the stresses of human movement, particularly on seams, won't compromise protective material" (McEntee, 1989).

More recently, the Defense Advanced Research Project Agency (DARPA) Autonomous Robot Manipulation (ARM) program has pursued the goal to develop software and hardware for an autonomous robot capable of using human tools and similar agile hands-on contact tasks via humanoid robotic arms, wrists, and hands.

The current publicly available iteration of the ARM robot platform has an overall humanlike representedness in its physical shape, which includes a head, face (with stereo cameras for "eyes"), a pan-tilt neck, two arms, hands (with force-torque tactile sensors), and a torso on a mobile base (DARPA, n.d.). Robert Mandelbaum, former Program Manager for the ARM initiative, was asked to give an example of the task that the ARM hardware and software under development would produce. He used Improvised Explosive Device (IED) disarmament as a specific example (Guizzo, 2010), indicating that Explosive Ordnance Disposal (EOD) work would be one area where ARM robots may be used.

In 2010, Vecna Technologies developed the Battlefield Extraction-Assist Robot (BEAR) for the U.S. Army (Gilbert and Beebe, 2010; Silverstein, 2010). BEAR is a 6'5" humanoid robot prototype that can lift up to 500 pounds, carry supplies or wounded soldiers, and is being investigated for other military applications.

Boston Dynamics has also developed Protection Ensemble Test Mannequin (PETMAN), a biped humanoid robot used for testing chemical-resistant apparel in the United States military (Shaker, 2011).

Figure 1.2   PETMAN

Source: Boston Dynamics, 2013.

In its final iteration, PETMAN will be "the shape and size of a standard human," according to Vice President of Engineering at Boston Dynamics, Robert Playter (Edwards, 2010). DARPA has also commissioned Boston Dynamics to develop the Atlas robot that is designed with a torso, two legs, and two arms (C. Brown, 2011; Edwards, 2010; Shaker, 2011). This robot has impressively nuanced physical capabilities and can walk upright using a bipedal heel-to-toe walking motion, maneuver sideways in order to move through narrow passages, and use its own forward motion to hurl or swing itself across gaps and between handholds.

Boston Dynamics states PETMAN was designed to test protective gear, and its humanlike shape allows repetitive motions of the robot to test for material issues. The robot also produces sweat and body heat like a human to determine whether a person can withstand heat and other related physiological stresses.

Figure 1.3   The Office of Naval Research-sponsored Shipboard Autonomous Firefighting Robot (SAFFiR) undergoes testing aboard the Naval Research Laboratory's ex-USS *Shadwell* located in Mobile, Alabama

Source: John F. Williams/U.S. Navy, 2014.

The Navy Center for Applied Research in Artificial Intelligence (NCRAI) is developing a bipedal, two-armed robot called Shipboard Autonomous Firefighting Robot, or SAFFiR, to assist sailors with damage control and inspection operations aboard naval vessels. SAFFiR is designed to move autonomously through a ship, naturally interact with people, and fight fires. In other words, it will carry out many of the dangerous firefighting tasks that are usually performed by humans (McKinney, 2012). According to a NCRAI press release, the plan is to enable SAFFiR to employ high-level reasoning ability and allow autonomous decision-making and mobility, making the robot a "team member" (McKinney, 2012, para. 4). Natural interaction, multimodal interfaces, and the ability to track the focus and attention of human team members will render SAFFiR a very humanlike robot in a military setting. The Navy further announced, "Like a sure-footed sailor, the robot will also be capable of walking in all directions, balancing in sea conditions, and traversing obstacles," (McKinney, 2012, para. 3).

Additional functionality will eventually include the robot's ability to understand and respond to gestures, such as human pointing and hand signals, and the robot will track the focus of attention of a human team leader. McKinney further states, "Where appropriate, natural language may also be incorporated" (2012, para. 4).

Other U.S. Navy humanoid robots, such as Octavia and Lucas, may become the next generation of SAFFiR, as their autonomy and social behaviors are integrated into the SAFFiR framework (Carroll, 2012; Webster, 2012).

Figure 1.4   Mobile, Dexterous, Social (MDS) robots like Octavia (pictured) are used by scientists in U.S. Naval Laboratory human-robot interaction research and to develop cognitive robotics systems

Source: U.S. Naval Research Laboratory, 2015

Lucas and Octavia can sense human commands and then decide upon and conduct a series of actions in response. To efficiently and effectively communicate with a human counterpart, the robots' behavior and appearance demonstrate their internal and emotional states. For example, a head tilt indicates the robot is considering a course of action. In addition, these robots can use speech to respond to people (Webster, 2012).

In 2004, through the Space and Naval Warfare Systems Center—San Diego's (SSC San Diego) research initiatives, came a proposed sophisticated system that would closely assist warfighters, "enabling a very synergistic teaming of human and machine capabilities" (Everett, Pacis, Kogut, Farrington, and Khurana, p. 2). The term *Warfighter's Associate* is based on this idea. It describes a two-fold concept in robotics: a (1) human-supervised platform that (2) employs a natural language interface and can understand and respond to high-level verbal commands, and is therefore semiautonomous. This model was developed in response to the emerging needs of the EOD units in Iraq and Afghanistan for robots with increased capabilities.

To illustrate their model, Everett et al. compared the idea to that of law-enforcement human-canine teams. They suggest that an anthropomorphic robot design might be better suited for some terrains and situations, while a Warfighter's Associate might be designed as a wheeled device for other scenarios (2004). This concept of a Warfighter's Associate would have a robotic embodiment that is still recognizable as non-living, but with some humanlike characteristics, like speech recognition. These humanlike traits might include the ability to interact in a human-robot team situation with natural language and a high degree of autonomy, exhibiting humanlike models of language and self-directed task-oriented behaviors.

Because of their usefulness in these ways, many robots used in defense will adopt more human and animal-like characteristics in future iterations. For example, in 2011, MGySgt. Carroll described plans for prospective new Explosive Ordnance Disposal robot grippers: "It'll have two fingers and a thumb, basically. We want some [haptic] feedback so we know how much pressure to apply for picking up intricate pieces of an IED," (Jean). Carroll's explanation illustrates why the properties of humanlike attributes such as fingers and the sensation of touch are useful in contexts such as EOD work.

While the level of anthropomorphism of different robot models will no doubt vary depending on their intended use, information about U.S. military goals to develop and incorporate robots that are more humanlike is publicly available. As cited in their 2004 report, Finkelstein and Albus (2003/2004) presented an EOD Mission Needs Statement published by the U.S. Army that plainly discussed EOD requirements for the concept of a humanoid robot:

> A need exists for a robotic platform that is capable of climbing narrow stairs, climbing ladders, opening doors/hatches, such as water towers, ships' holds, or roofs. The humanoid robot would be capable of climbing both ship and land-based ladders. A humanoid robot would alleviate a need for the robot to be light

for transportation, since it would be able to stow itself into an EOD response vehicle. A humanoid robot would also be capable of emplacing a disrupter tool or x-ray rather than the current methodology of mounting the disrupter on the tracked or wheeled robot. (p. 106)

The Mission Needs Statement goes on to outline disadvantages of the wheeled and tracked (tanklike) maneuvering robotic systems used at the time, citing the robots' weight (weighing hundreds or thousands of pounds), which complicates transport, slows down movement, and causes problems related to scaling different types of terrain. This disadvantage in turn prevents the disruption of devices on rooftops or similar tall human-made structures. This statement also suggests investigating Current Off-The-Shelf (COTS) humanoid robots that could be modified for EOD use. There is also a movement toward developing robot squad members that are meant to replace human soldiers, at least as remotely controlled physical stand-ins for some situations (Dyess, Winstead, and Golson, 2011). Humanoid robot design is clearly being examined as a potential design choice for widespread defense use.

Appliqué, which refers to add-on armor, is another method of increasing the adaptability of things, from people to machines. One advantage of adding armor to an existing resource, such as a tank, is to customize an off-the-shelf product to respond to a specific threat. Human operators will be able to wear a version of appliqué or gear that is a type of robotic exoskeleton. These complex robotic wearable systems can improve individual performance, as well as act as protective armor.

The Army's proposed Tactical Assault Light Operator Suit (TALOS) will be a system of wearable tools deemed most useful to a soldier. While wearing it, strength is increased for tasks such as lifting, and it will likely help with endurance as well. TALOS is planned to incorporate responsive material fitted with sensors to monitor body temperature, heart rate, and hydration levels. Some aspects of medical triage may even be carried out in real time for the wearer, like the application of injectable foam blood-coagulating products applied to a soldier's wounds by the suit itself.

Just by wearing an exoskeleton, a soldier's everyday experience with the world around them will change dramatically with extended senses and abilities. With the widespread and instant availability of wearable and implanted technologies, the experience of war will be very different, very quickly, once again. Enhanced performance tools will allow a soldier to seamlessly view their environment using thermal imaging, night vision, smart optics, and Remotely Piloted Vehicle (RPV) feeds concurrently. Any soldier, regardless of training or expertise, may be able to communicate in a foreign language using effortless simultaneous translation. These rich tools will lead to complicated changes for the suit-wearers and those they interact with, whether colleague or enemy or ally.

Elsewhere in this book, there are more detailed discussions of the military as an organization and its impact on EOD work, individuals, team dynamics, and robot design. Unlike many strictly hierarchical military working environments, EOD personnel are uniquely situated within the military for many reasons, not least of which because they are trained so highly to work and communicate as teams. In EOD work, unlike other groups within the military, Team Leaders frequently ask for opinions from all team members, including junior or newer members. EODs—as personnel are frequently referred to—are formally and informally trained to share information through ongoing communication, which is recognized as a critical part of their decision-making processes. Each teammate may have only partial knowledge relevant to solving the problem, different competences and skills, and potentially differing beliefs about the state of the tasks at hand.

Continuous communication is also critical in the event a team member is injured or killed while disposing of unexploded ordnance. In the event of a casualty or death, remaining team members must understand why each person in their unit chooses and plans to do something, so that they may troubleshoot when possible if communication is lost and learn from the outcomes of each unique situation. Although one or two team members may be tasked with operating or maintaining the robot regularly while other members have different expected duties, all team members are exposed to the robots' use as a tool and are frequently in close proximity to the robot.

Given the nature of this type of closely interactive team work, coupled with the utility and frequent use of robots every day in these dangerous team contexts, it is perhaps not surprising some soldiers begin to feel an affection for their non-humanoid robots, similar to what they might feel for a pet (Singer, 2009).

Storytelling

Humans are storytellers. People speak to one another about their experiences and the meanings that these experiences have in their lives. Although the very word *story* may connote something fictional, people also create and tell their own real story every day. Worldwide, all cultures and societies create and disseminate stories about their past, present, and their view of the future. Through stories, people create a shared history. They use stories to organize thoughts, elicit emotion, and teach others in their community how to live and behave and what their expectations should be. Storytelling and personal narratives can be used to clarify positions, provide vicarious experiences, incite actions, share experiences and ideas, reduce conflicts, and act as touchstones for people sharing common cultures.

Narrative or *storytelling* are the ideas and related actions of networked connections between a series of events that are retold to an audience that generally agrees upon how the story is understood. It is not a list of events, but describes events around a unified subject or theme. In addition to story's core elements of character and sequenced actions, a narrative offers a scaffolding condition, or a cause that suggests how the events are linked in an identifiable way.

A *plot*, then, characterizes these narrative events and guides the audience through a thematically connected time line of these episodes as the connections created by the storyteller. Stories use plot to tie story events together, despite the possible temporal shift between separate events or events that may be presented in a nonchronological order. Thus, a story presents a coherent whole in its telling, tying together incidents in a way that makes sense using the plot as a lens of understanding. Using this paradigm, a narrative is about its conveyed and understood meanings, and it presents a series of thematic actions in a chronological framework. The actions do not have to follow a linear time frame, and the event order and the actions described are carefully chosen by the storyteller as a collection, connected by the plot structure. The sequence of events in a story does not have to be strictly chronological, although it can be. The order of events can include digressions, foreshadowings, and flashbacks. In fact, because narrative is propelled by events, the goal is not essentially analytical or critical. Yet stories can contain moral lessons, such as in genres like fables, fairy tales, or folktales. Therefore, plot structure presents the order in which an audience becomes aware of what transpires in the story.

Because different storytellers may present their own versions of the same experienced or imagined events using their own beliefs or narrative preferences, searching for consistency across versions is an exercise in connecting the identifiable themes, even with elements such as events, characters, or plot transformed in the retelling. This explanation of storytelling is not meant to be an exhaustive theory of narrative, but rather to clarify its characteristics from other ways of communicating. This basic frame emphasizes some of the significant components of narrative, such as the storyteller, the audience, events or episodes, temporal frameworks, themes, and meanings.

Versions of truth and beliefs are conveyed in stories. While a story itself may not prove a thing occurred or existed, it does demonstrate how something might come to be. They are causal links in a successful story that the audience recognizes, when those same causes may not be so identifiable in discursive communication. People use narrative and discourse every day to make sense of their world. Life experiences are conveyed as plots with characters, and explain ways of thinking by the things the storyteller chooses to tell, the medium of telling/understanding, how stories are told, and in what order events are presented. Audience understanding is based on shared knowledge about cultural meanings and expectations. The audience participates actively in the narrative interaction, placing primacy and attention on details salient to their own understanding, beliefs, expectations, and preferences, and interpreting narrative accordingly. Thus, successful narrative uses a logic structure that at its core relies on this shared understanding between storyteller and audience. Because of these characteristics of narrative, its qualities offer a unique way to understand ways of experiencing life, inner motivations, beliefs, and values when analyzing a story and its reception. The formal structures of language, genre, medium, characters, and shared understandings of these factors express truths, what is valued (or not), what matters (or does not), and connections between ideas, people, things, and events that may seem otherwise logically unrelated.

Stories and narratives do not necessarily have to be created with the idea of entertaining an audience as the goal. Storytelling is a model of communication, and as such contains the very human emotional qualities of social interaction integrated into communication. Storytelling can be cathartic for the creator as well as the audience, the engineer and the product, the designer and the user, the maker and the market. As an exercise in creativity, a medium for sharing ideas with colleagues, a persuasive action to incite actions, or simply a way to share news, narrative is a message that reflects its creators' thoughts.

There is value in storytelling and sharing experiences, whether the stories are fictional and metaphorical or based on real experiences. Storytelling is a platform for discussion about beliefs, can influence expectations, and presents a canvas for interpretation by everyone who participates in the storytelling. In everyday lives, people use real, personal stories to share memories, and to do so, often communicate their experiences episodically. These episodic stories often begin with familiar signaling phrases such as, "I remember when ... " or, "This morning, I ... ." These stories are generated by individuals to organize and frame the meanings of things they have experienced. In other words, internal stories help people give subjective meaning to themselves, form how they think about others, and inform their expectations about the world around them.

Storytelling about a real event can open channels of thoughts, beliefs, feelings, and communication for the teller and the listeners, informing and affecting people at emotional and behavioral levels. The opportunity to recount one's past to a compassionate listener—especially when the storyteller can integrate what they experienced in past events into present, everyday life—often leads to the telling and reception of deeply personal stories. Personal narratives, similar to other forms of storytelling, are not simply reporting *who*, *what*, *where*, and *when*. Narrative also attempts to frame *why* something happened, an interpretation of events presented by the storyteller. The storyteller makes sense of the action for the audience by grounding it in their own way of thinking.

By casual observation, clearly some of the ways people participate in the stories they tell are by the very acts of creation, presenting, and sharing. Less obvious to the naked eye is how people decide what stories to share with others, and how they choose to share them. Perhaps even more challenging for an outsider is to discover others' internal, individual experiences when they participate in the stories they hear and see.

Personal experiences with even the same story can change over time, as when one watches a favorite movie or reads a book again. People will sometimes revisit a story because it is familiar and enjoyable. However, familiar stories can elicit different meanings, focus, or insights from the same person through repeated *participation* because the story audience's perspectives evolve. Whether shared as a book, movie, advertisement, folk tale, or any other form, stories continually take on new meanings. People are not just influenced by storytelling in a straightforward manner, but rather iteratively mold a version of the story to fit their lens of interpretation.

Still, the circle of influence between stories and culture does not end so neatly. A storyteller can be responsive to an audience and so change the way they tell the story. In addition, the roles of "storyteller" and "audience" are fluid; even in an everyday conversation, it is natural to take turns in these parts. The associations between stories and people are dynamic interactions, as are the processes of building a story.

The idea of cooperatively generating personal stories may seem counterintuitive at first. Yet people collaborate on narratives all the time. Consider the situation of several friends hiking in a wooded setting. One friend points out that an unleashed dog runs across the trail ahead and comments about it, just as another hiker notices it, too. The first friend's comment may alert the other to the situation, or perhaps it confirms what they believe they saw.

A split second later, a third companion on the same hike informs the others that the "dog" they saw was actually a wolf. What is that person's authority for knowing? In essence, this question is asked at some level every time another person's point of view is different from his or her own. Did the second friend have a better vantage point for seeing the animal? Is that hiker very familiar with dogs or wolves? Alternatively, is that friend perhaps prone to exaggeration? Possibly, based on the beliefs about one friend's knowledge of wolves, some members of the group decide to reevaluate the moment of the "dog" running past. Nevertheless, the first friend maintains it was a dog. After further consideration and discussion, the majority of the group agrees the idea of a wolf rather than a stray dog in this setting makes more sense. As a result, they become more alert to their surroundings. They stifle an instinct to call after the animal that was first regarded as a stray, possibly domesticated animal. Still, the subjectivity about the story has changed. Additionally, individual and group behavior modified as most of the members became more vigilant and aware of their surroundings based on the revised belief they saw a wolf. Even the version of events and how the story is told in the future may be changed. This exchange demonstrates how descriptions of events often involve a negotiation of interpretations.

No two people will ever understand a moment or a story in the same way. Stories are not exercises in objectivity. Yet acknowledging subjectivity of experience does not mean it is undeserving of analysis; on the contrary, it is worthy because it is a first-person description. Furthermore, the primacy of first-person experience often reveals what was significant to the individual when the event is shared later, even if the version of the tale evolves over time or is adapted for different audiences. It is the very subjectivity of a narrative—and its influences—that can contain truths about the storytellers.

Returning to the earlier example, what about the truth? Was it a dog or a wolf? Which friend was right? Although there is an ultimate truth about whether everyone witnessed a dog or a wolf run past, each person's version of individual interpretation of the moment is still their own truth, and so has its own validity as experience; what is very real is each person's subjective experience and the emotions and behaviors tied to that experience. When determining immediate personal-threat levels, it may be crucial to know in the moment if it was really a stray dog or a wolf in the path. Yet the negotiated narrative process in that story also affected immediate behavior changes without this concrete knowledge of the "truth" revealed. In the dog-versus-wolf example is an example of the rapidity of role-changing between storyteller and audience, the fluidity of objectivity, the influence of other interpretations on stories, and how interpretation can affect behaviors. Therefore all of these processes should be examined with equal fervor when pursuing the mystery of the identity of the animal on the path. Only investigating all of these things together leads to the philosophical metaphysical meanings of the phenomenon, thus gaining insights into the causes, as well as outcomes.

Science and Fiction

Humans are organically connected to experiences in the world. After all, each human's nervous system, their very physicality, positions them uniquely and therefore individualizes their relationships with experiences. According to a legendary cinema history anecdote, in 1895 the first audience for the 50-second film *L'arrivée d'un train en gare de La Ciotat* (Lumière and Lumière) was so startled by the moving pictures of a train heading toward them that they fled from their theater seats. Regardless of the veracity of this story, the tale conveys that the first movie audiences were unprepared for the illusion of movies. However, after over a century of moviegoing, audiences are now prepared for a movie based on their prior cinema experiences. They do not run from oncoming movie trains on the screen because there is an expectation the images are just that, and thus do not represent immediate physical harm. Human relationships with a story and a technology, as with movies, can change over time as experiences with the technology inform evolved ways of interacting. In addition, collective ideas about subjects and concepts change over time, and stories change to reflect the evolving ideas of the larger society.

A modern horror movie audience will go into a theater with the expectation of seeing something that will make them scared, titillated, surprised, and a slew of other emotions that are normally categorized as negative feelings in our everyday lives. Why do some people enjoy these elaborately staged shows when they understand as modern moviegoers that what is seen on-screen is not real? They know the movie dangers are not really in the theater, able to reach out and harm them. Then why do situations on a screen sometimes create very real behavioral reactions, such as someone's startled jump in their movie seat during a surprising moment on film, if there is an understanding that it is a movie and not a real threat to real, physical being? The reason audiences can react physically is in some ways connected to the same questions about why they laugh or cry when a film is emotionally moving. In some ways and on some occasions, people interact with the movie, a non-living thing, as if it is a real situation.

When watching a threatening movie scene, as opposed to a neutral one, parts of the brain are activated and aroused, alerting a sense of personal danger (Straube, 2010). Reading a scary book can have similar effects, causing physical and emotional anxiety (Cantor, 2004). Stories are evocative and hold human attention because, at a certain level, when immersed in a good story people actively participate and share the experience of the plot. There is an understanding the story is only that, a story. Yet if stimulated by a story, people can react viscerally.

Throughout this book, there is an approach to human-robot interaction as an exercise in cultural studies. It is something that is discussed through a lens of understanding that these interactions are already embedded in our culture. In order to more deeply investigate the ways in which our culture contructs and disseminates knowledge about human-robot interactions, it is important to look at fictional storytelling and our larger shared culture, as well as to explore individual narratives about self. Analyzing the properties of stories as narrative develops a basis for understanding parts of different cultural systems.

To begin to make sense of stories, recognize that there are common properties in how to define what makes a complete narrative structure. Narrative accounts, fictional or real, have two fundamental properties. First, they focus on someone's intentional state, or their values, beliefs, opinions, and views. Second, they recount how these states lead to behaviors or activities. In fiction, the turning point of a story is highlighted so the audience can more clearly understand the connections between these two elements, the internal motivations and goals and the external actions. In reality-based narratives, discovering the connection between the two sides of self—internal and external—is less easily pinned down. Closely examining the texts and contexts of real stories for patterns of expression and action reveal the core truths, which in turn can lead to insights about how people see themselves and the world around them.

Works of fiction, folk storytelling, and popular culture have always influenced people's expectations of robot appearance, behavior, and purpose (Kaplan, 2004). Stories contribute to the making of our own experiences, but also create common understandings in our culture, acting as touchstones for social relationships. Myths of artificial life in different forms interacting with humans have been told since antiquity in many cultures.

The word *robot* actually has its common roots in fiction. The Czech word *robota* was the inspiration for author Karel Čapek's (1920) term for the humanlike artificial agents that make up some of the key characters in his story *RUR* (*Rossum's Universal Robots*). In *RUR*, Čapek's worker robots are portrayed as tireless and uncomplaining workers until they are imbued with emotion; once emotion is installed in *Čapek*'s robots an android uprising kills the humans and robots rule the world. This story is almost certainly a modern template for many narratives about human-robot interactions that followed.

Fictional concepts about artificial human life have been omnipresent around the world for thousands of years in mythology and folklore, such as Prometheus's clay men (Hyginus, c. 900/1960), Hephaestus's creation of Pandora (idem), and

Pygmalion's Galatea statue come to life (Ovid, A.D. 8/2009). There are stories about a Chinese automaton so humanlike that its creator, Yan Shi, could only defend it as artificial by disassembling it in front of the king (Chen, 1996). In the Middle Ages, legends began of robot guardians in Lokapannatti (Strong, 1994), and of giant golem protectors forged from clay in Jewish folklore (Goldsmith, 1981; Idel, 1990).

Modern-era popular stories about robots or artificial life combine elements of plausible technology with unanticipated results in order to create a titillating character or dramatic story plot. But for every lovable robot portrayed in films such as *Star Wars* (Kurtz and Lucas, 1977) or *Wall-E* (Morris and Stanton, 2008), there are many tales told that focus on dangerous robots such as Gort from *The Day the Earth Stood Still* (Blaustein and Wise, 1951), *Blade Runner*'s replicant-robot assassins (Scott, 1982), and the near-indestructible *Terminator* (Hurd and Cameron, 1984). Even the sympathetic Tin Man's crude robot character in the 1939 classic movie *The Wizard of Oz* (LeRoy and Fleming) pines for a heart because the tinsmith that built him forgot to give him one.

This record of conflicting themes in popular stories of robots reflects the apparently universal longstanding cultural confusion about what to do with artificial life, and presents examples of ways people might interact with artificial life. Like movies, storytelling through folktales—oral tradition, books, and songs—add to our model of understanding things that are otherwise not easily understood. The traditional tales surrounding the creation of golems are particularly relevant, as they became root fables for many subsequent stories about automatons like robots.

Consider the historical context of Jewish people as a European minority during the Middle Ages. Frequently under both random and systematic physical attacks in ghettos and villages, the idea of a massive, magical protector like a golem was a popular story no doubt at least partly because it positioned Jewish people as empowered with possibilities when, in reality, everyday life was very dangerous. In other words, golems made for good stories in escapist and aspirational respects. In fact, the provocative set of philosophical questions golems pose about the meaning of life, creation, and Otherness are appealing enough that they continue to inspire modern fiction.

Traditionally, golems are described as artificial men created out of clay or dust and animated with the "breath of bones," or *ruah*, also referred to as the "animal soul." The distinction between the states of *animation* and *living* are clear in these stories, and a golem is not considered one of God's creations. Rather, these clay men are something human-made, a sort of auspicious tool constructed from magic, purpose, dirt, and spring water. This set of unlikely ingredients is no less believable than those mentioned in other popular stories about imbuing something human-made with a sort of animation, or imitation of life.

Golems are described as humanlike in shape and physically large, but not intelligent, and without any autonomy. As automatons, golems must follow the orders of their creator without internal critical thinking mechanisms that are recognizably human, and they are completely obedient. Yet the primary purpose of the process of a golem creation was intended to be a spiritual exercise. Rabbis might create a golem as a way to cleave to God and perhaps gain further insight into God's plan. The golem story can also be interpreted as presenting the golem creator with an opportunity to conjure a doppelganger, a copy, or lesser version of self. The golem is then a vehicle for the rabbi to understand his own nature through observation, leading to a thoughtful personal change on a lifelong path to personal redemption.

However, other golem stories tell how the creatures so effectively served their creator that they were assigned repetitive household tasks like drawing water from a well. From these examples of the golem's tireless obedience, it made practical sense to use the efficient creation as a tool or force to protect villages from attack. According to the lore, when golems were used for menial service or physical protection, these improper uses of the golem led to chaos, usually by the golem morphing into an uncontrollable thing. Golems, like other human creations leading to disorder, are also part of a common motif of stories centering on themes related to technology. People use stories like this to explore the outcomes of their curiosity, and the potential ramifications of experimenting with the unknown at such a high level.

Film, television, books, and computer-mediated stories are critical pieces of society where myth-making and meaning-making emerge and combine with what people know from real experiences. Contemporary anxieties associated with technology are bound up with many cautionary stories about humans overstepping their abilities when creating artificial life, such as Shelley's (1994) Frankenstein creature or the robot child in the film *AI* (Kennedy, Spielberg, and Curtis, 2001). Consider that in turn, these story themes and metaphors contribute to contemporary attitudes toward scientific advancements (Nerlich et al. 2001).

An example of this interplay between storytelling and advances in science are Isaac Asimov's Three Laws of Robotics (1950). The Three Laws have become a framework—appropriately or not—for some to understand how to even approach developing an intellectual, moral, and ethical rubric for robot behaviors. Without expanding into the debate about the application of this fictional device to science, it is worth noting that these Laws are not only a pervasive cultural touchstone among science and science fiction fans, but have been disseminated with a certain amount of influence into popular culture. Murphy and Woods claim, "The Three Laws have been so successfully inculcated into the public consciousness through entertainment that they now appear to shape society's expectations about how robots should act around humans" (2009). While it is impossible to say how pervasive its influence might be, there is merit to the claim that Asimov's rule went far as to shape individual expectations of what would constitute a normal human-robot interaction, or at least how the robot should behave. Furthermore, media frequently refer to human-robot interaction in terms of the Three Laws, popularizing this set of ideas. Currently, human-robot interactions are not yet common or everyday experiences for a large part of the population. However, as robot use expands from specialized situations such as in militarized spaces, our narratives and expectations about the technology will no doubt shift, too. Over time, the underlying ideas in fictional accounts like Asimov's exploration of what makes us human may be explored even more deeply, or stories may turn to a more external focus, critically reflecting on what makes humans work with robots the way we do.

Science fiction is a means of confronting current understandings with the complex implications, probabilities, and possibilities of new human-technological interactions. In the 1972 movie *Silent Running* (Trumbull), one scene illustrates a very salient interaction example, relevant to this discussion on design. In the movie, Freeman Lowell is the only human aboard a spacecraft. For much of the film, Lowell's only remaining companions are three vaguely humanoid robots named Drone 1, Drone 2, and Drone 3. The drones are small in stature and only humanlike in their bipedal mode of movement and most basic morphology; they communicate non-verbally, and their functionality is defined by their interactions as service robots. They are not designed with an intent to be socially interactive. Yet when Lowell is alone with the robots for a prolonged period, he begins to emotionally bond with them. One of the first ways Lowell mentally elevates the drones to something lifelike is to rename them Huey, Dewey, and Louie. In a twist of irony, even the reference for those names comes from other stories of anthropomorphism, associated as they are in American culture with animated talking-duck characters.

In this film, the robot design was clearly set with built-in boundaries for interaction, with the robots presented with machinelike form and limited intelligence appropriate for their tasks on the ship. These drones did not have natural language speaking ability, although they apparently have a natural language understanding. Yet when left alone with no human companionship, Lowell names the drones—in effect imbuing them with more emotional meaning than purely functional objects or tools could carry. In an early scene, Lowell plays poker with his human coworkers. In a parallel scene, after Lowell is isolated from human contact, he teaches the robots how to play cards, and they act as poker-playing companions for him. Throughout the movie, Lowell speaks for the robots, articulating and narrating their interactions and providing what he imagines their speech might be in a faux back-and-forth social transaction. Clearly, the robots become companions for him in his loneliness. The movie presents examples of how the audience might imagine a human operator violating the robots' design intent as purely service models. When Lowell is faced with an opportunity to use the robots in unanticipated ways, he does so, quickly turning to them for social interaction and building on their limited built-in humanlike cues with his imagination. It is a poignant example of science fiction's examination of the roles robots might play in people's lives.

Mythmaking

A myth is a story that has wide cultural acceptance and is often associated with sacred characteristics, such as communicating fundamental truths about life.

Moreover, it is a story in which ordinary people and objects may have the power of gods or heroes. When talking about mythology, there is a distinction between the sources of the mythmaking story and other ways of thinking, such as those that might be deemed based in *logical*, *rational*, or even *scientific* thought. Myth creation is based in human perception, emotions, and familiar models of interaction. However, it would be simplistic and inaccurate to say that because mythopoeic thought is based in emotion and feelings it is not necessarily logical, or that this state acts solely as an operation that occurs prior to logical or scientific thinking.

These stories present multiple explanations and alternatives for phenomena where the details are not otherwise easily understood, and are entertaining, often presented with a humorous edge.

Myths may be integrated into different social circles of culture, from an individual level to the family, organization, or larger society. They capture a culture's truths and reflect its concerns, reinforcing aspects of a society through the effective use of archetypical symbols in storytelling that capture the audience's imagination and stir questions and thoughts. Myths are not simply for entertainment, but present versions of truth and beliefs. They reflect some of the psychological, sociological, and metaphysical ideas that are important to the people within that culture.

In the field of anthropology, *mythopoeic thought* refers to a stage of human thought that is distinctly different from scientific thought (Frankfort, Frankfort, Jacobsen, and Irwin, 1977). Specifically, mythopoeic thought generally refers to a period of a culture's ways of thinking that is part of a social evolutionary process that leads to seeking scientifically rooted concepts. Myths emerge from understanding natural phenomenon through different means than a scientific view. Myth- and science-based systems are two very different ways of perceiving and interacting with the world. Some cultures operate around beliefs largely rooted in mythology, while others are more science-centered. One method of understanding is not necessarily intrinsically better than the other, and it is not obligatory to evaluate them as methods that serve at cross-purposes. Both concepts are equally significant to those who rely on science or myth as a lens to view life. Yet it is important to distinguish between the two sets of viewpoints when analyzing the germination of beliefs, truths, and emotions associated with things.

One of the crucial differences between the two is that is unfair to judge either using the opposing lens to view it. To say it is less sophisticated to find meaning in myths is akin to dismissing emotions in general, and their very real part of the human experience. In all human everyday lives, the realties influenced by these modes of thought are blended, fluid, and operating simultaneously, affecting individual experiences, memories, and past and future behaviors.

Tied to the idea of mythmaking is the premise of *speculative thought*. Using this model, speculative thought takes two forms, *rational* and *mythopoeic* thought. *Rational thought* uses logic, verifiable facts, and peer discussion to propel ideas forward. *Mythopoeic thought* finds order and meaning through the use of myth, metaphor, storytelling, and narrative. Historically, these theories were used to understand specific eras of ancient and modern cultures, presenting mythopoeic thought as more concrete and humanizing, while a scientific approach is viewed as impersonal and abstract. There is a contrast between approaches, but not an inherent exclusivity between these ways of thinking. Often people use a combination of these methods when approaching a new and complex problem. Therefore, consistently underestimating one means of understanding or the other does the analysis of human interactions a disservice. In fact, mythopoeic thought could be an early part of the understanding process for any group when they encounter a new thing, experience, or way of thinking.

One of the characteristics of mythopoeic thinking is that it assigns all visual and auditory stimuli a biomorphic cause. This projection of ideas on to a problem allows someone to understand how to interact with the thing or situation, triggering vital behaviors, including defensive or other protective measures. It is part of a way to assess the power, intelligence, agency, and friendliness of an unrecognized or unfamiliar other. Using these beliefs, every unknown object is initially imbued with a state of consciousness, as a living thing.

Rational thought tends to presume there are thinking, aware objects, and then inanimate objects on the other end of a life spectrum. Thus, objects are examined with tests, analysis, and observation that are rooted in the aforementioned belief system. Objects are things, whereas people are understood to have experiences, emotions, and feelings. Mythopoeic thinking means there are no objects in the world that do not have lifelike and humanlike attributes. In order to understand other humans, people employ empathy to identify with these living others. In mythopoeic view, everything is understood through empathy, not universal laws and generalizations applied as law. Determination of how to treat an *other*, living or non-living, is therefore understood by human actions, behaviors, and attitudes.

Both of these modes of interpretation are part of many people's everyday lives to varying degrees. As an example, science fiction is frequently used to explain the unexplainable. In the not-too-distant past, humans did not interact with robots. Books, movies, and television depictions of human interaction with robots proposed ideas that audiences recognized as possible outcomes of human-robot contact, usually portraying robots based on a human-centered way of existing, with humanlike morphology, speech, and interactions. As actual robots become integrated into a modern world, previous ideas presented as fictional may seem naive, silly, campy, or simply inaccurate when compared to the new reality. Yet the stories serve purpose beyond entertainment. These stories provide metaphorical ways of comprehending new things, like robots, using a human-human metaphor in order to make them comprehensible. Personifying robots is one way of understanding something for which there are no universal laws yet. Using myth and storytelling to understand these new things provides some order and paradigm for cultural thinking and exploration of ideas about robots if they are fictionally portrayed as humanlike individuals.

For the purpose of further cultural analysis, there can be a distinction made between a story understood by both the storyteller and audience as fictional (e.g., a science fiction movie), and one rooted in a storyteller's version of reality (e.g., a news story). In the case of humans interacting with robots, these two categories of narratives are sometimes blurred as people become accustomed to the idea of real robots—even interacting with them in real spaces—yet do not have much, if any, personal experience with an actual robot. This historical time in human-robot interactions reflects the initial cultural norming process, where humans are discovering every aspect of how they think they should live and work with robots. Therefore, there may be especially active reflections between fiction and real stories in this era of discovery. The results, determined as the models for how society and individuals should act with robots and live with them, will be a dynamic tension between informal and individual decisions and the formal policies and laws developed to govern human-robot interactions.

It is perhaps easier to understand how the real world can inspire fiction, and be the springboard for fantasy. To what extent do the stories people invent impact the "real" world? What, if any, is the role of fiction in co-constructing the technological practices of the robotic designers, engineers, and scientists?

There are several ways to position fictional stories and their influence on everyday life. Storytelling can be studied as a vehicle for analysis, as it is in one form in this book. During a prescribed interpretation of specific cultural phenomena, such as human-robot interactions, there is an examination of the messages about norms, identity, and understanding presented by the fictional stories as a reflection of what exists or occurs in real life. Often in tandem, this sort of detailed examination of storytelling also examines who is telling the story, the context in which the story is told, the medium by which it is conveyed, and other situational variables.

Widely recognized fiction, science fiction, and other popular culture references are frequently used to report on robots in everyday conversations, news stories, and even in formal wording by policymakers. Headlines that use the Terminator to discuss real robots and news stories about real robots that refer to Asimov's fictional Three Laws use these well-known stories as cultural touchstones, appealing to a shared understanding of the implications of these allusions. These symbols are easily understood shorthand references to present complicated information in a familiar way, using a commonly understood fictional interaction model to convey a representation of what might happen in reality. When using a fictional reference as a descriptor, the reference also becomes a frame for the model of understanding, and may act as a positive or negative influencer, depending on how it is used as a rhetorical device.

News reports of emotional connections between troops and robots are already becoming part of our collective cultural narrative about how humans have—or might—treat robots (Garreau, 2007; Rose, 2011). In many news stories about robots used in defense, journalists have chosen popular fictional robots with negative associations—usually those commonly viewed as faulty harbingers of uncontrollable doom—as a point of reference. Using this strategy to signal common understanding through a symbol with destructive connotations sets a tone for the human expectations of the robot (or robots in general), as well as the tone of how the article frames the issues. This circle of cultural influence may also conflate issues in some possible human-robot interactions, exaggerate ideas based on fictional examples, and rely on inaccurate tropes in order to add an entertaining aspect to what can otherwise sometimes be a difficult topic to discuss in few words: our minimal current understanding of the complexities of human-robot interactions. In this way, fictional examples may negatively reinforce people's understanding of what they think they know, or what they imagine working with real robots might be like. Clearly, one obvious advantage to using references to popular stories is they serve as shared cultural reference, a way to communicate complicated ideas by using a model of interaction that is recognized in terms of these shared meanings in familiar symbols. If someone speaks about the Terminator robot, it invokes a set of expectations in the listener about how a robot might look, act, or what its tasks or capabilities are.

Related to EOD work, news stories typically describe a type of affection from soldier to robot, often reporting the robot as named by the team members with fondness. It is possible these associations stem from the relatively small size of most robots that EOD personnel currently use, the fact that these robots are not typically weaponized, and that they are designed low to the ground, without clear humanlike attributes. On occasion, the words of people interviewed form the basis of the fictional comparison, as in this example: "Troops call their robots Johnny 5 if they're dependable," (Komarow, 2005). Narratives such as these have also painted a picture of robots as mascots or lifelike addendums to the human teams, and help form the public's perception of some types of military robots, as well as help form initial expectations about interacting with these robots.

Scientists, engineers, makers, and roboticists are also people involved in social systems. Thus, fiction and stories can also inspire scientific ideas. Many robot makers and designers will cite their own childhood-rooted fandom of particular robots from movies or books, and connect that to their real life aspirations. Programmer and designer Paulius Liekis is a fan of robotics and 3D printing. He turned to one of his favorite movies, the science fiction anime *Ghost in the Shell* (Oshii, 1995) for inspiration. Relying on screen shots from the film to guide and inspire him, Liekis produced a 3D model of the movie's T08A2/R3000 spider tank. "If you look at the tank in a movie it has a shape of a robot, but movement is more similar to a dog or some other live animal. You can feel emotion from its pose, you can see that it's angry or that it's trying to protect itself," explained Liekis (Millsaps, 2015).

Alternatively, it appears to be an intuitive proposition that science fiction writers are also influenced by real-world technology. Screenwriter Jeff Vintar, who counts the adaptation of Asimov's *I, Robot* short story for the big screen among his many writing credits, explains his own way of thinking about the genre: "Science fiction can be anything, a love story, a monster tale, an epic adventure. The only thing that makes it 'science fiction' is some use of future science, some extrapolation, of what might be" (personal communication, 2015). Vintar discussed the engaging aspects of both fiction and real science had for him from an early age:

As a kid who fell in love with science fiction? I can only guess why. But what do children dream of if not far off places, and tomorrow? They long to be grown up. To do things they can't do now … Before the internet, when I was just starting out, I would spend hours inside bookstores poring through science books, on robots, on space and time, anything that caught my eye. Science fiction does ask you to invent, and the most practical place to look for inspiration is in the real world of science. *The Age of Spiritual Machines*, *The Cyborg Handbook*, and other volumes were piled high on my desk all the years I wrote an original script called *Hardwired*, which then became the screenplay for *I, Robot*. A science-fiction writer who does not keep at least one eye open to what's really happening in the various disciplines would be like a man who decided to turn off his senses, so convinced that he could create everything with his own imagination that he no longer needed to see, hear, smell, or taste the world. That might work for a while. But he may open his eyes again to find that the world passed him by. Don't get me wrong: a science-fiction writer is just as likely to be surprised by what tomorrow brings as anyone—and because we believe we're on top of things, odds are we'll be even more dumbfounded—but we have a responsibility to at least try.

Pepper, a social robot designed for consumer home use and developed by Adleberaan for Softbank Mobile, was released in 2015. Softbank Corp. CEO Masayoshi Son has said the inspiration for Pepper began with his memories of Astro Boy from childhood (Kambayashi, 2015). Astro Boy is a popular animated Japanese robot character created in the early 1950s by Japanese animator Osamu Tezuka. In his origin story, Astro Boy was endowed by his creator with a mechanical heart and humanlike emotions. Thus, similarly, Pepper has been endowed with a "heart" and can behave with a semblance of humanlike emotions, as well as identify and recognize emotional cues observed from interacting with people. In fact, Pepper can even appear to cry, with lights around his eyes imitating a welling up of tears.

In the political arena, scientific experts advocating for the minimization or prohibition of the creation and use of lethal autonomous weapons (LAWS) and lethal autonomous robots (LARS) are concerned about keeping a discussion open about the ethical considerations of building these systems, and related policies. Various groups have lobbied humanitarian and government agencies for disabling, disarmament, and disuse of these systems. Initially, these lobbyists faced skepticism about the nature of the requests to begin a dialogue on these issues due to the resistance to accepting the reality of these autonomous weapons issues, with opponents using the term *science fiction* to marginalize the issues. Similar fictional references are used in the coverage of these issues by those in the media—and are now sometimes employed by the anti-autonomous weapons groups themselves—to signal other broadly recognized meanings, such as the perceived potential impending crisis of an enormous magnitude if these systems are widely used. As stated in *Foreign Policy* (McCormick, 2014), "The killer robot has been a science-fiction staple for decades, but rapid advances in artificial intelligence may soon usher in the era of lethal autonomous machines ... A growing chorus of critics think machines shouldn't be licensed to kill ... But which would you put your money on, the UN or Skynet?"

No active member of society escapes cultural influences. People use stories as touchstones, models of interaction, to present new ideas, and as paradigms of expectations about human-robot interactions. These ideas are often naturally integrated into their work, in turn perpetuating some concepts rooted in fiction, influencing others in the connectedness of culture. Journalists, roboticists, policymakers, law enforcement, and military personnel tell their own stories, and are participatory social actors in the exchange of beliefs and expectations about living with robots. Again, as part of culture, the science fiction references, like their real-life story counterparts, will evolve and change over time, as technology and society change.

Chapter 2
Explosive Ordnance Disposal Stories

The springboard for this book was a study designed to increase the understanding of everyday robot interactions in the U.S. military. In order to streamline this enormous overarching goal into a focused and feasible study, research concentrated on individuals within the military who interacted with robots every day. At the time this work began, Explosive Ordnance Disposal (EOD) personnel were one of the most easily identified groups of people who interacted with robots almost daily at some point in their service. Like any cultural group, Explosive Ordnance Disposal personnel consciously and unconsciously collaboratively create a collection of shared knowledge, artifacts, and meaning associated with these things. Due to the nature of their service responsibilities and their common experience of job-related robot use, for many EOD personnel, their experience places robots as a significant part of this system. By focusing on EOD personnel as an initial way to explore experiences with robots, the study examined structures, workings, and social origins of military human-robot interactions in one of the first groups to use robots as part of their common tool kit.

The data in this case consisted of both questionnaire results and personal narratives of soldiers. To obtain these narratives, a short one-on-one interview was conducted with each soldier. Interviews are a data-collection method, but also a structured social interaction between a researcher and a participant who is identified as a source of information, as characterized in this situation by EOD professionals. If during the interview the interviewee is probed about beliefs, attitudes, and experiences, the information is, as it stands, highly contextualized and therefore considered local knowledge. When an appropriate sample of participants have been interviewed and their responses are studied for emerging patterns in interview responses, the resulting analyses can help pinpoint what areas of interest to probe further, or with a large enough data set, produce generalizable knowledge claims.

The analysis process for this study looked at the ways social phenomena are created between the larger institution (the U.S. military) and troops' individual experiences in the world, from recruitment throughout various stages of their careers. This type of research places an emphasis on the significance of culture and context to understand what occurs in society, and constructing new knowledge about situations using this way of understanding phenomena. The ideas of *social constructivism* are rooted in specific notions about human-perceived reality, knowledge, and learning. Using that premise, reality is something constructed through human activity; members of a society create properties of the world around them (Palinscar, 1998; Kukla, 2000). Thus, reality is something that does not exist prior to its social invention. Similarly, knowledge is also a human product, and is socially and culturally constructed (Gredler, 1997; Prat and Floden, 1994).

Because individuals generate meaning through interactions with one another and with their environment, viewed this way, learning is also a social process. Thus, learning is not something that occurs only within an individual, but is an active process shaped by people's interactions with external forces (Vygotsky, 1986; Palinscar, 1998). When individuals are engaged in social activities, meaningful learning will take place. Engaging in collaborative learning is therefore a dynamic process of developing a communal "social world" (Palinscar and Herrenkohl, 2002, p. 28).

Another way to view this is to consider the whole, dynamic interconnectedness between people and their social relationships and environments (Bredo, 1994; Gredler, 1997). As a result of these collaborative processes, learning does not take place separately from an environment, but rather as part of it through ongoing interactions. If the roles and responsibilities of an individual change, social and transactional relationships among group members change (Bredo, 1994; Gredler, 1997; Palinscar and Herrenkohl, 2002) as well as their situated perspectives.

One characteristic of a social group is it is a collection of individuals whose interactions are based on common interests and assumptions. Ways of seeing the world and understanding associations between ideas, experience, and things develop and evolve through negotiation among group members (Gredler, 1997; Prawat and Floden, 1994). This construction of meaning and subsequent shared understanding between individuals that forms a basis for their communication is termed *intersubjectivity* (Rogoff, 1990). One example of this shared negotiated meaning and understanding is language, which is a system of tools and signs understood within a group and used to mediate ongoing learning (T. Duffy and Cunningham, 1996; Palinscar, 1998). All successful communication and interactions require using and understanding the socially agreed-upon ideas social patterns (Ernest, 1998; Vygotsky, 1986). Consequently, intersubjectivity is the basis of communication and helps people expand their understanding of new information and activities among group members (Rogoff, 1990).

Intersubjectivity also applies to our understanding and use of tools, as well as how and why they are created. Any time a tool is developed and becomes part of common use, it transforms the tasks of the user. In turn, this changes how the user interacts with everything else in their environment. A tool as robust as a robot is laden with cultural expectations about its characteristics and use by designer, user, and the interaction observer. The very goal of incorporating robot technology into everyday interactions is to change the human roles and tasks; by successfully doing so, the human tasks and roles will change.

A robot's role in the military is often positioned as a tool to provide support to soldiers, and one that is even an acceptable substitute for a human troop in some environments and situations, albeit with human intervention or guidance (Finkelstein and Albus, 2003/2004; Lin, Bekey, and Abney, 2008; Magnuson, 2009). EOD robots are an example of technology as an intersubjectively understood part of the activity in an environment. The robots' role is one that can support tasks and create new activities, and therefore provides changing operator focus and their subsequent knowledge about situations.

According to social systems theory, a *system* is defined by a boundary between itself and its larger situated environment, dividing it from a complex external world (Viskovatoff, 1999). Therefore, the interior of a system is a zone of reduced complexity. Communication within a system functions via group system member selection of only a limited amount of all information available outside. The criterion according to which system information is selected and processed is *meaning*. However, systems comprise both physical and observable behaviors as well as subjective and less concretely quantified internal and individual motivations, preferences, emotions, and intentions (Viskovatoff, 1999).

In this case, the EOD *microsystem* (Bronfenbrenner, 1979), or immediate workplace surroundings of the EOD individuals and their work peers, is depicted with an overview of the people, operating environment, and everyday tasks in Chapter 4. Further, the *mesosystem*, or relations between the different microsystems, is explained through a description of the connection between EOD common school training and shared experiences beyond the workplace in the larger EOD culture and military organization.

The theoretical base of this social systems approach to data aligns well with the traditional breakdown of topics in human-robot interaction (HRI) research, such as task, environment, and social modeling (Burke et al., 2004). Consequently, the original study situated EOD individuals within the larger systems of their organization, operating environments, and job-related tasks in order to illuminate the conditions and environments in which EOD human-robot interactions occur.

PART II
Metaphors

The field of human-robot interaction is an interdisciplinary domain. Looking to other disciplines' models of understanding complex human-nonhuman emotional connections is part of the ongoing exchange of ideas between specific areas of study. This multidirectional interchange of ideas also illustrates how there can be an increased richness of knowledge across domains.

Linguistic metaphors are used in language every day as a type of communication shorthand, a way of using words outside their normal understanding in order to convey observations between different concepts. Consequently, metaphors are not just the words spoken or written, but are fixed in thoughts, ideas, beliefs, and preconceptions. It is a way to map things across knowledge domains and connect abstract ideas (Lakoff, 1992), and is a process based on experiences, both as individuals and as part of larger social systems. Similar to metaphors, the concept of *metonymy*—where words and expressions are used as a stand-in or shorthand for another thing mapped to similar associations—refers to a contiguous process, stemming directly from a stimulus. However, metonymy usually implies a direct correlation between a thing and someone's experience, while a metaphor is commonly considered as a way of understanding or explaining something that has been contemplated or believed, not necessarily requiring a straightforward experience.

Two models of human-other interaction are presented in this chapter: human-military working dog (MWD) and human-object attachment. These examples are included as more than models of human interaction with non-human things, but as metaphors for how people might view robots now, and potentially sources of insight into how interactions with robots might change over time. In theory building, models are types of metaphors as well. In a way, it is still a type of shorthand. Models are used to investigate new phenomenon through a comparison to something similar that is already understood using existing theory.

One of the challenges of choosing which models to examine is that using an existing set of concepts to explore unknown concepts is not a neat process. A model is, by its nature, only partially like something else. The key to effective use of modeling when developing theory is choosing models that are a good fit, and ensuring that the models used do not contradict what is already known about the phenomenon being studied. The models chosen as a means of understanding other things must also pass the tests of internal and external validity, and be rooted in logically comparable systems. However, the very nature of a metaphor or a model is that the two compared things do not match perfectly. Rather, it is this tension between what is known and the unknown that ultimately mediates any understanding of similarities and differences between the two. As such, theoretical models are not substitutes for new ideas, but rather scaffold the construction of new knowledge by providing a framework for building theory.

Chapter 3
Our Emotional Engines

When discussing collaborative work environments, there are several relevant emotion-based concepts that need to be addressed because they affect human decision-making in teamwork or cooperative work scenarios between people: (1) bonding, (2) team cohesion, (3) trust, and (4) attachment. The emotional elements of these important team characteristics are a source of ongoing investigation and debate.

Group *cohesion*—the linking of people via bonding in a group—incorporates the social and task-oriented factors of bonding with the addition of a perceived group unity and emotional aspect (Forsyth, 2010; Johns et al., 1984). In group cohesiveness, the social facet is based on the relationship that members feel toward other group members and toward their group as a whole. This type of cohesiveness can be both a formal and informal type of social structure (Kirke, 2009). Group cohesiveness is *formal* in that it exists within the more global military hierarchy as a specific subgroup: Explosive Ordnance Disposal.

This structure exists also *informally* as a group of ideas and norms within EOD that may not be explicitly named, but that form a set of conventions and expectations for behavior and actions. A relevant example for EOD work (Wong et al., 2003) when discussing social cohesion cites shared combat trauma as a strong condition for bonding between soldiers. A system of robust informal bonds among group members contributes to the overall cohesion, as does the more formal convention of a clearly structured set of goals (Kirke, 2009). It is important to note that cohesiveness is not something that is constant, but is dynamic and an ongoing negotiation involving individuals within a group and is largely dependent on *operating structure*, or the shared cooperation of a task, goal, or mission (Kirke, 2009).

Bonding can be described as the process of developing a strong interpersonal relationship with others. Human-human *bonding* is more than merely liking another person. The process of bonding refers to an interpersonal relationship developed over time. At this stage of technological development, robots cannot return affection, affinity, empathy, or other complex human emotions. Therefore, any potential model of bonding in this discussion will refer to a one-way model, or human-to-robot. The process of human-human bonding can occur over social and task-related components (Eisenberg, 2007), including task cohesiveness, which refers to the degree to which group members share collective goals and labor together to meet these goals.

Bonding can include affection and trust as parts of its definition. The bonding process reduces negative stress and is a mutual, shared activity between people.

Attachment theory (Bowlby, 1973; 1980; 1982) poses the claim that the more time a person spends in close proximity to another person, emotional bonds will be strengthened through an *affectional* or *emotional bond.*

Emotional bonds are something humans instinctually seek (Bowlby and Ainsworth, 1989). Historically, much of the research in the area of attachment has been applied to parent-child relationships or couples. Affectional ties manifest over a period of time, and are persistent, not fleeting. This theory of the attachment process and bonding is applied to specific aspects of human relationships. When people are hurt emotionally or physically or even perceive a threat to their selves, attachment is viewed as a system of behaviors used to remedy these situations.

The emotional characteristics of this bond can have related behavioral outcomes, such as when an individual wishes to maintain proximity or contact with the person with whom she has an affectional tie. Furthermore, separation from the figures of attachment can cause varying levels of distress. When viewed as a way to form bonds with trusted others, then attachment is a used as a way to regulate stress by helping to define the scope of those who are deemed physically and emotionally safe and supported. Another criterion of emotional attachment includes a person's need to seek security and comfort in the bonded relationship.

Similarly, trust develops based on prior experiences and interactions with someone or something, and help forms expectations about how these others are likely to behave in the future. Developing trust can also involve seeing others as personally motivated by sincere care and concern to protect another's welfare. Trust is also an essential requirement of a functional relationship between humans to ensure that collective goals and outcomes will be effectively worked on together. Hancock, Billings, and Schaefer (2011) define trust as "the reliance by an agent that actions prejudicial to their well-being will not be undertaken by influential others" (p. 24). Using this definition, a human's trust in a robot's behavior and reliability is necessary for effective human-robot interaction to transpire. Trust also results from learning about situations and others, and is therefore an embedded process in social systems, and thus it is a means to manage expectations about relationships and the environment. Consequently, trust is a part of all social relationships, as well as a means of predicting the future for and by individuals. It is a critical factor in human relationships because it influences interaction results via attitudes, behaviors, and perceptions. In the case of EOD work or similar potentially life-threatening human-robot cooperation, issues associated with operator trust toward the robot are of particular concern because it is necessary for a person to rely on a robot for the safety and welfare of themselves and others.

Past interactions are used to create a framework of understanding about what the other person or thing is likely to do in specific situations. Moreover, this information is used to create increasingly predictive models of what this other is likely to do on a regular basis. Therefore, trust is also an important predictor of how people will behave toward others. One way humans choose to trust someone (or something) is by determining if there is more benefit to be gained by trusting in this interaction than is risked. Because trust is rooted in a process of understanding and expectations of another person, there is a common belief that a decision to trust is purely rational. Trust simplifies and reduces the complex set of expectations that used to predict how people and things will behave, work, or perform. Yet trust is influenced equally by assumptions, feelings, judgments, and attitudes about the other person's perceived associations with social categories. Even with this invisible and evolving internal set of heuristics used to determine a level of trust in another, behaviors related to trusting or not trusting a teammate or tool have observable outcomes.

The need to trust people stems from an instinct to predict the behaviors of others in order to keep ourselves safe. Because it is impossible to predict how other people will act in every situation, there is always a risk for adverse consequences when trusting another person. Therefore, the act of trust is ongoing, dynamic, and can change over time, increasing or decreasing in valence. Without the ability to trust others, people would have to be perpetually on guard and hypervigilant, a condition almost impossible to sustain emotionally or physically and still function well.

In risky conditions, the level of vulnerability can depend on a person's successful interdependence with others. The uncertainty of a situation increases interdependence in order to increase the chances for surviving with minimal risk. Trusting other people leaves one vulnerable to other types of risk, such as the loss of the other that was depended upon. Cooperation is different from trust. Someone may prefer to cooperate with people they trust, but trust is not necessary for cooperation.

A military context can appear to be the ultimate environment to explore trust issues, as the nature of a military organization and its tasks presents some of the greatest conditions for individual and group risk and vulnerability. There is an ongoing need for effective cooperation among its members, and widespread interdependence between colleagues is integrated into everyday life. From members' initial training throughout their military career, they are taught the effectiveness of efficient teamwork. Conversely, they are also aware of the potential costs of misplaced trust, ineffective cooperation, and high interdependence on others, and the increased risk in situations already deemed dangerous. Trust—as well as confidence—in teammates and job tools is critical in order to increase the likelihood of survival. Therefore, understanding what factors influence the development of trust has great value.

Ideas about trust often come from understanding the relationships between people who one interacts with every day, or what is sometimes referred to as *person-based trust*. Furthermore, expectations and beliefs about people are reinforced or negated as familiarity is extended to them over time. However, another type of trust can form between people without any direct interaction between them: *category-based trust* (Krämer, 1999). Category-based trust can occur when the other person is a member of a group or category of people that has been trustworthy previously. Category-based trust can also transpire from a shared affiliation, such as being a member of a group with another person. This type of trust can also be rooted in the training and experience associated with the jobs or roles of others. Again, using what is learned previously from similar situations helps to assess characteristics of the unknown, such as whether to trust a colleague, and if so, under what conditions. Consequently, person-based trust builds categories on a model of understanding, and is another tool to simplify one's environment.

When meeting a stranger, one immediately begins to use cues such as voice tone, physical stature, and facial expressions in order to try to understand the other's motives. Although it is over time conceptual ideas are refined about categories and definitions of group members are formed, these associations are still used immediately—in concert with all other information available— to form a shorthand for understanding. The same information is used to identify if someone is a member of one's own group, too. This stronger identification with someone else, such as with an identified shared group affiliation can increase the level of trust for that person. One reason this may occur is that when someone feels connected by a social category, the joint mission and motivation expands in scope beyond simply assessing one's own personal environment to include the safety of other members of the group.

Human-machine trust models posit that the characteristics used to determine trust in a machine include predictability, dependability, competence, reliability, and responsibility (Muir and Moray, 1996; Rempel, Holmes, and Zanna, 1985). In this case, *predictability* applies to the machine's behaviors, the operator's ability to predict machine behaviors, and the soundness of the machine systems.

As with human-human paradigms, this model of trust is an iterative one. Developing and maintaining trust in a machine will only take place if the machine carries out the desired behaviors effectively and repeatedly, reinforcing positive expectations about machine performance over time. In order to maintain trust, ongoing validations are sought to match positive expectations of a machine. For the machine operator, this means they need to adjust their own behaviors according to their evaluation of the machine.

Another factor that may affect a user's level of machine trust is an individual's trust in the machine's designers. One relevant example of this trust could be someone's familiarity with a robot company and their reputation for stable models. Associating positive traits from the company with the machine developers' abilities and, in turn, applying that faith to the resulting machine may establish trust even before a person has used a system. In the same way, the actual machine design is dependent on the creator's perceptions about the relative capabilities and needs of the operator.

An operator's psychological state while working with the machine also affects trust level. The operator's self-confidence and general predisposition to trust automation can influence behaviors. Working with semiautonomous robots, this could be demonstrated as the operator's choice of when to override any autonomous machine feature with their human decision-making.

Commonalities between different trust models indicate some consistency between how trust is constructed across systems. However, this assumption is not as straightforward as it may appear. The human-human and human-machine models of trust may have similar dynamics, but society is still determining what model will emerge as a framework for human-robot trust.

In the military, from the early days of training throughout a career, the concept of working with others in a cohesive unit is pervasive. Yet the idea of working with a robot as an active part of a cohesive unit may seem unusual because *cohesion* is not necessarily something thought of as important in many human-tool situations. One definition of horizontal cohesion (Stewart, 1988) includes the components of peer bonding, technical proficiency, teamwork, trust, respect, and friendship. Yet this definition is fluid; as the nature of war changes, things that create cohesion change. For example, modern teams that work together in teleoperated environments but did not necessarily train or live together must create cohesion in order to be effective.

Consistent messages reported from EOD people working with robots (Roderick, 2010) demonstrate that human operators currently trust EOD machinelike robots, facilitating decisions about sending in robots *in lieu* of humans in dangerous situations. This finding leads to questions about whether the conditions of this trust change if the robot design resembles a living thing, or is perceived as something lifelike.

It is not surprising when robots are referred to as living things; sometimes it is because people use human-human interaction cues mapped on to human-robot interaction, and other times it is because robots seem to exist in a new social space. How EOD robots are presented to personnel is another topic worth examining because it bootstraps the expectations of functions and roles on the team or within the unit. In a 2010 U.S. Army recruitment video, robots are positioned as team members (Inside look at 89 D—Explosive Ordinance Disposal Specialist). The video shows an EOD team in various aspects of their jobs, with team members speaking to the camera. "We run in three-man teams. [*Gestures at QinetiQ North America's TALON robot*]. This is our fourth member. We can send this guy out and he does the dangerous stuff for us," Sgt. Dean states in the Army video. Later, in the same video, Staff Sgt. Mitchell proclaims, "Our robot driver is Dean. So we just call this robot mini-Dean." Whether these are scripted comments or genuine remarks are less important than the fact that this robot is referred to as a team member and as an extension of the user (Sgt. Dean).

This scenario may influence any viewer about how to interact with a robot—primarily in a social or interdependent manner—as opposed to positioning it as an inanimate tool like a tank or rifle. From the Armed Forces projected standpoint, the robot does stand in for the EOD personnel as a sort of troop doppelganger or an extension of their physical self. This purposeful message to the public that robots act as tools to extend and save human lives has many latent implications in the interpretations of this representation of technology. It is true that robots are a valuable tool to EOD personnel and can save lives by acting out tasks in dangerous scenarios. However, the inaccurate representation of the robots as possessing characteristics such as humanlike membership in the team positions the robots in a social context that arguably simultaneously legitimizes and minimizes the use of this type of technology in warfare by imbuing it with value as more than a tool. Therefore, this presentation could make the idea of this technology more palatable for potential recruits as well as to other audiences, reassuring them that robots seamlessly mitigate risk to human life and do so in an understandable, humanlike way.

There is no question that judicious use of semiautonomous robots in dangerous situations offers some operational and tactical advantages. Robots can minimize the risk to human life by taking over dangerous tasks previously done by humans; are impervious to chemical and biological weapons; increase the endurance, agility, and strength of users or operators; and may be designed to influence human operator situation assessment. Yet, while a robot that is perceived as lifelike may be extremely effective at specific tasks, it can still elicit unpredictable emotions in the users, depending on the situation.

Emotions are a significant component of a functioning human and are tied closely to their actions and reactions. How people appraise situations and others can result in distress, relief, anticipation, hope, frustration, pride, dislike, affection, contempt, surprise, fear, and an infinite list of fluctuating states humans use for self-reflection, to act in circumstances, and employ to assess people and things in the world.

Freud, in a psychoanalytic literary analysis about gothic horror, claimed that a sense of the *uncanny* is what elicits a sense of horror, and occurs when a person or a thing appears simultaneously strange and familiar (*The Uncanny*, 1919). Furthermore, Freud explained that things were especially uncanny when they were "capable of independent activity." Mori (1970/2012) developed a graphical illustration to accompany his theory about human emotions and interactions with humanlike robots called the *uncanny valley*. The word *valley* refers to a dip in the y-axis (familiarity) of Mori's proposed graph showing the positivity of human reaction as a function of a robot's lifelikeness. Mori's theory states that the more humanlike a robot is in its appearance and movement, the more positive and empathetic a human being's emotional response to the robot will be. However, a point on the x-axis (humanlikeness) of the graph that occurs when the entity is almost indistinguishable from a live human pushes the human response to strong repulsion.

Mori also posits that the reverse holds true; as robot appearance becomes less distinguishable from a human being, the emotional response becomes positive once more and approaches human-to-human empathy levels. Thus, the "dipped" area of repulsed response aroused by a robot that has appearance and motion between a "maybe-human" and "perceived-human" entity is referred to as the *uncanny valley*. One implication of this theory of uncannyness is that humans may feel robots that appear too human are unsettling, perhaps because they are in an unrecognizable category of thing that resembles something organic, but clearly is not. Therefore, any initial user perception of uncannyness in a robot has the potential to distract the operator at some level.

An important aspect of defense work is *stress*, or rather any variable of it that disrupts the normal functioning of an individual. There are two different psychological models of stress commonly addressed: stimulus-based and response-based. Yet both of these models are largely rooted in examining environmental stimuli and can ignore individual traits and experiences. A more holistic approach to deep understanding of human emotion and behaviors tied to stress also requires an investigation of circumstances, situation, and context. Consequently, some experts on stress maintain there are no absolute psychological stressors—what may cause stress for one individual may have comparatively little emotional or behavioral impact on another individual (Stokes and Kite, 2001).

Better suited to this type of study is the concept of the *transactional model*, which posits stress is a dynamic interaction between an individual and their environment, with an emphasis on the role of the individual's situation appraisal in shaping their responses (Lazarus, 1966; Lazarus and Folkman, 1984; McGrath, 1976). One definition of stress is a mismatch between "individuals' perceptions of the demands of the task or situation and their perceptions of the resources for coping with them" (Stokes and Kite, p. 116). Keeping in mind that description, stress does not always have to be characterized as something negative.

Stress—which depends upon emotions, environment, individual personal characteristics, and inter- and intrapersonal interactions—affects people's cognitive and decision-making abilities, affecting the outcome of simple tasks and complex operations (Lazarus, 2006). However, investigating stress by itself does not do much to tell us how individuals adapt, short- or long-term. Investigating stress in tandem with emotions and attachment theories tells us more about how people appraise and cope with their situations. A better understanding of coping practices can then lead to discovering practical solutions to the complex social relationships between human-robot team members.

Although stress is often defined in terms of negative feelings and associations, certain conditions of stress can actually help survival by triggering hormones or other physiological reactions necessary for homeostasis. Individuals are commonly aware of stressors when they feel confused, frustrated, or challenged. People working in defense, space exploration, and humanitarian relief efforts are often positioned in situations where personal safety is in jeopardy, life-dependent social connections are formed with teammates, and the everyday undertaking of a work-related task can take place in a dangerous or hostile environment. Reactions to negative stress range from mild anxiety to debilitating mental disorders, and organizations responsible for people in these conditions must be aware of how it affects the well-being of individuals within the larger units.

Regardless of the cause, stress may affect many crucial tasks and behaviors that are crucial parts of EOD (or other) work, such as attentional processing, task management, working memory, and decision-making (Staal, 2004). However, the focus of this early stage of inquiry into EOD-robot interactions is not to make generalizable claims that any specific tasks, situation, or environments are stressors for EOD personnel. Rather, this work explores the human-robot relationship to discover if this dynamic includes variables that introduce stressors and, if so, begin to define and determine what aspects of human-robot interaction cause negative stress.

Chapter 4
Meaningful Connections with Non-human Things

Morally Significant Forms of Attachment

Some types of attachment may be considered morally significant. People take for granted that a type of attachment exists for things, such as possessions. If these possessions are ruined or taken away, the owner may become upset or distressed, and they would be less so if they did not feel a sense of moderate attachment to these things. Attachment to other people is important, and the inability to develop or maintain attachments to others is considered problematic, an indicator of abnormal behavior, and even a psychological problem to be resolved. There are times when attachment to beliefs such as religion or cultural identity are considered morally legitimate, and that actions threatening these attachments are moral wrongs. Clearly, these forms of attachment have different roles in people's sense of identity and psychological welfare. People may become attached to possessions because of their aesthetic value or usefulness, but a religious, cultural, or national identity is more closely tied to a sense of self.

By defining attachment in terms of its importance, certain violations of these constructs are considered morally wrong. For example, to limit someone's religious freedoms is regarded as a moral wrong because it strikes at the connections between identity, self, and culture.

However, some forms of attachment are not considered morally significant. Even a sincere attachment to a fictional character is usually not viewed as serious. In general, people do not believe that when a fictional character is harmed there is a real moral wrong, even if those attached to that character become upset. In fact, often people who are strongly attached to fictional characters are viewed as outside the norm, and it is thought that they should mitigate their attachment and recognize the attachment is not "real," or that the attachment will inhibit an ability to effectively function as part of society.

Is Attachment to a Robot Morally Significant?

On the surface, attachment to a robot appears similar to attachment to a fictional character or imaginary friend. Robots are real things that interact with people in their world, but are sometimes endowed with unrealistic social constructions or humanlike qualities by people interacting with them. One question that emerges from this line of thinking is whether people should be actively discouraged from becoming attached to robots. After all, robots are not living things. Yet robots are different from fictional characters and imaginary friends because they exist in the world. They can be customized to reflect an owner's preferences and to interact with people, other robots, and their environment, and they can even be forms of self-expression for the people who control them. In these ways, robots can become more closely tied to an individual's personal identity than an imaginary entity. Robot attachment is then not easy to dismiss as illegitimate or even morally insignificant without rejecting other, similar forms of attachment.

Another concern about attachment to robots is the fear of loss or other distress if the robot is harmed or otherwise disabled, causing emotional repercussions for the people emotionally engaged with the robot. It is true that detachment may reduce stress associated with a robot's demise or injury. Yet attachment is related to enjoyment and engaging with an other. Discouraging attachment to a robot can undermine an individual's motivations to work with or engage the robot in activities because the interaction is not pleasurable, or as pleasurable as expected. While attachment can supplement and reinforce a sense of trust or loyalty in a human-human model, it is possible encouraging detachment in human-robot interactions can undermine people's reasons for interacting regularly with a robot. Moreover, an attempt to dissuade human attachment may introduce stress, creating a cognitive impediment counterintuitive in the nonintuitive form of interaction. Still, an emotional detachment from robots is possible, and may even be the desirable condition in some contexts.

Military Working Dogs

One relationship model between humans and nonhumans that can be looked to for possible insights into human-robot relationships is that of human-animal teamwork. As companion animals and pets, animals are related to in ways that parallel the social exchange humans have with other humans. There is a mutual recognition between pet and owner of co-participation in the interaction, a shared use of imagination and expectations created in order to understand the other, and a joint adjustment of position and behaviors based on how each understands the other's perspective. Everyday social interactions with pets are generally perceived as mutually rewarding social experiences. Pets are rewarded for good behaviors, and in turn, owners bask in the positive feedback received from the animals as their successful caretakers.

Pets and companion animals also inform and change the way behaviors are modified based on constructs of how the animal is perceived. Companion animals are regarded as sensitive, reciprocating, and integrated socially into people's lives and families. From this position, the pet acts as a member of the group, and is treated as such. Individuals sometimes go so far as to label themselves as a "dog person" or "cat person," declaring an allegiance to a favorite animal group, even if this claim is purposefully exaggerated and said with self-awareness. Pets become a defining part of people.

The significance of human-animal bonding to the situation of EOD-robot work is twofold:

1. Research suggests that human trust and bonding with robots may have some emotional parallels to human-animal bonds (Billings et al., 2012).
2. How the military classifies and regards working animals is a potential paradigm for how robots are classified now and in the future. This official classification, in turn, affects troops by explicitly positioning the robots' social role within teams.

Human-animal partnerships are unique and can benefit people emotionally, physically, and cognitively (Levy, 2007; Wilson, 1994). Although Military Working Dogs (MWD) work with their soldier handlers closely, the U.S. Defense Department currently assigns Military Working Dogs—like robots—the classification of "equipment" (*Robotic Systems Joint Project Office: Unmanned Ground Systems Roadmap Addendum*, 2012). This categorization of MWD comes from the necessary defaulting between the only two choices the military currently assigns assets: humanpower or equipment (Cullins, 2011). However, soldiers forge strong emotional bonds with these canines that are acting as part of their team regardless of the formal "equipment" classification. Because of the strong bonds soldiers forge with these animals, there is a passionate MWD advocacy movement proposing to change the military classification for working canines from *equipment* to *manpower* (or a third, as yet undesignated category) in order to initiate and clarify policies for prolonged care and maintenance of the dogs after retirement (Cullins, 2011; Rizzo, 2012).

The way society regards animals can appear to be contradictory. The animals or pets in our home are considered family, but outside the home our relationships with animals change (e.g., work, feral, and animals raised for food). In other words, some animals are treated almost like people while others are used as tools, where people have otherwise modified or justified their relationship to the animal's status as a living thing. Breeding is actively manipulated, picking and choosing animals to be better food, hunters, or companions. Outside of a military setting, it has been suggested a way to categorize all legal policy regarding animals might be as "living property" (Favre, 2010).

A military context is not an unusual context for human-animal joint effort. Animals have been used in warfare throughout history, and the story of dogs in war goes back to antiquity, as many forces recognized the value of trained dogs on the battlefield. Canine assistants have been used in defense for transport, weapons detection, communications, and as comfort to soldiers on and off the battlefields. As of 2012, there were a reported 2,700 dogs serving with the U.S. military worldwide, with 600 of those active in designated war zones (2012).

In World War II, via the Dogs for Defense program, troops worked with civilian pets that were volunteered by their owners for military service, and then subsequently trained and integrated into military specialties such as explosives detection. One example from the World War II era of human-MWD bonding is exemplified in a letter written by Marine PFC Wachtsletter to a dog's owners, to inform them the dog had died in service (National Public Radio, 2012). Wachtsletter wrote of Tubby, the Working Military Dog, "He behaved like a true Marine at all times and didn't even whimper when he died. We've buried him at the Marine Cemetery along with the other real heroes of this campaign … He has a cross with his name and rank. He's a corporal."

Dogs for Defense ended in 1945 because of the many logistical problems involved with borrowing civilian dogs and retraining them to be integrated back into their original families, post-specialized military canine training. However, MWD owned completely by the military are no less likely to bond with their human trainers and handlers. Lackland Air Force Base spokesperson Gerry Proctor (Rizzo, 2012, para. 18) states, "A handler would never speak of their dog as a piece of equipment. The dog is their partner. You can walk away from a damaged tank, but not your dog. Never."

In fact, the story of one handler who tried unsuccessfully to adopt his retired MWD partner provided the basis for Robby's Law, signed in 2000 by President Clinton (Sesana, 2013). Robby's Law permits the adoption of military working dogs by their handlers, law enforcement agencies, and civilians who apply to the adoption program. Furthermore, as of 2000, the Secretary of Defense must submit an annual report to Congress accounting for all MWD adopted under the program, those on the adoption waiting list, and those euthanized. The report must specify the reason for euthanasia instead of adoption for all dogs that are euthanized.

The roles and tasks of MWD are varied and include sentry duty, airborne, search and rescue, and bomb detection. In essence, the dogs are used for dangerous and repetitive tasks otherwise performed by humans (or, more recently, by machines). The dogs are highly trained and often assigned to one handler, who then works with a dog for years at a time, including periods of deployment. It is not uncommon for a handler to adopt their primary MWD after the dog's retirement.

A significant part of the canine-handler partnership is established during training, when the handler must demonstrate their ability to develop a relationship with their assigned MWD. Cpl. Johanna Robbins said, "When you deploy [with your dog], it takes being man's best friend to a whole new level. Your life depends on them as much as their life depends on you" (Hurtado, 2014).

Research using zoomorphic robots has demonstrated that in some conditions and circumstances, imbuing robots with animal-like characteristics may support effective human-robot interactions (Arkin, 2005), and using a robot with animal-like traits can also affect human-operator perceptions about the robot's intelligence and abilities (Bartneck, Reichenbach, and Carpenter, 2006; Bartneck, Reichenbach, and Carpenter, 2008). Therefore, there is an interesting set of potential emotional dilemmas for troops working closely with these robots, and projecting any sort of trust, bonding, or attachment to an inorganic thing that can be destroyed, put in danger, replaced, left behind, abandoned, or treated as any other piece of military equipment. The very things that can prime someone emotionally to become attached to another human—such as the need to feel safe or comforted when threatened with harm, or when they are separated from others they care for—are contexts that are exemplified from moderate to sometimes extreme ways within military work.

The LS3 robot is quadruped, and its movements resemble those of a pack animal, similar to a mule. It is used primarily as a logistical tool, supporting load-bearing tasks and acting as another means of transporting supplies in terrains difficult for wheeled machines. An LS3 can be programmed to follow troops, similar to a dog or other animal. Moreover, while the robot's role is to assist the troops, on occasion the troops must take care of the robot, too. In addition to needing any general maintenance, the robot can sometimes lose its footing and tip over, requiring the help of at least one troop. While any tool needs regular care in order to work well, the nature of the human-robot interdependence sometimes replicates situations similar to human-animal care. As these robots are being tested and incorporated into units, some news has reported aspects of affectional ties by troops toward the machines, such as naming the robot, and comparisons made by troops characterizing the relationship to the robot as one they would have "like a dog" (Dietz, 2014).

Product Attachment

One way to define product attachment is as an emotional bond to a specific product. Like human-human attachment, this relationship implies that a strong tie exists for the individual to something else, but in this case, the other thing is a consumer product. There is a common understanding that children become attached to a comforting doll, adults can feel pride in getting the latest gadget from their favorite company, or a soldier might name their rifle and a pilot their plane. People better care of these things that have meaning to them, even with the clear understanding these meaningful things are "only" non-living objects. Over time, because of recurring pleasurable experiences while using or interacting with the product, the object becomes meaningful. If that specific object were to be lost, taken away, or destroyed, an individual could experience emotional distress in varying degrees.

Battarbee and Mattelmaki (2002) describe three categories of human-product attachment that closely parallel themes pertaining to human-human attachment. The first category is "Meaningful Tool," in which human-product attachment occurs because the object serves as a symbol for a highly valued capability. The second category is "Meaningful Association," in which human-product attachment resides within an object's association with a valued cultural meaning, such as membership to a specific social group. The third category in this model reveals the complexities involved in human-robot attachment. This category proposes that individuals may treat products as "Living Objects," a condition in which the product "is a companion that has been with a person for so long that it is perceived as having personality, soul, character, is loved and cared for" (p. 4). Other factors that can influence product attachment include a sense of self-expression associated with the product, or whether the object is connected to the idea of making the individual self unique in some way. Memories related to the product, which imply a longitudinal relationship, can also affect feelings and behaviors about that thing.

Emotion and attachment are intertwined. Human behaviors indicate a level of commitment to objects that hold these meanings for them. If someone feels strongly attached to a product, they are more likely to be careful with the object, repair it as necessary, and even postpone its replacement when its intended use is impeded. Note that while all of these factors are relevant to stimulating attachment to products, the degree to which applied product design strategies affect these human behaviors varies. For example, it will bolster the likelihood of user attachment to develop a product so it is customizable or able to be personalized by the operator. Yet people may choose to customize objects, even complex systems like robots, without this design intent in place through relatively simple acts, such as assigning it a unique name, or through more complex processes, such as jury-rigging a product to make it exceptionally useful to the user.

Based on current understanding of human-robot interaction, it makes sense to look to the research in human-computer interaction (HCI), especially the theories of why and when people ascribe human qualities like gender and politeness to even disembodied machines (Reeves and Nass, 1996; Nass and Moon, 2000). Many people affectionately name their cars, boats, dolls, or even weapons. The findings of Sung, Guo, Grinter and Christensen (2007) showed just over two-thirds of their study participants named their Roomba, a vacuum robot used in the home, and many assigned the vacuum robot a gender, referring to the it as "he" or "she."

Reeves and Nass's (1996) "computers as social actors" (CASA) theory proposes humans unconsciously ascribe agency, personality, and intentionality to computer-mediated technologies. Similarly, the combination of robot appearance, behaviors, roles, and user-projected human intentionality creates a complex mixture of attachment-related responses for users of robots in which the drive to respond to a robot as if it were human is at odds with the realization that the robot is a machine.

One of the central themes in human-computer interaction research has been *how* and *to what degree* people personify computers and attribute human-like qualities to computer functions. Particular emphasis has been placed on the social aspects of human responses to different communication technologies. This research has demonstrated that the relationship between humans and media technologies is fundamentally social, meaning that social dynamics and constraints usually associated with interpersonal relationships apply to mediated interactions as well (Reeves and Nass, 1996).

In 2012, Michael Kolb conducted research specifically examining the dynamics between humans and robots that work together in a high-stress military combat environment. Based on the results of a Web-based survey completed by

746 (soldier) participants, Kolb compared and contrasted human-robot bonds to human-to-human bonds that are formed under the same stressful combat conditions. Among other findings, Kolb came to the conclusions that (1) the findings did not prove conclusively that working with robots in combat contexts increases bonds between humans and robots, but the high stress circumstances *may* contribute to the initial formation of bonds, and (2) the idea of *emotional attachment* of humans to robots in military contexts was not proven in this study. However, Kolb also acknowledges that the idea of human-robot attachment in this sort of military human-robot scenario "could change in the future as robotic development advances" (2012, p. 80). This work measures feedback from soldiers in a recent timeframe, and demonstrates the self-reported beliefs and attitudes in that time frame and among a wide variety of jobs within the military. Yet the process of attachment is not an immediate one. Attachment and bonding research about soldiers and robots needs to be reevaluated iteratively, over time.

There has been promising research on the development and uses for human-robot socialness, conducted with a spectrum of robots that vary in intelligence, behaviors, appearance, abilities, and autonomy and studied in a variety of contexts (Breazeal and Scassalleti, 1999; Breazeal, 2003; Fincannon, Barnes, Murphy, and Riddle, 2004; Fong, Nourbakhsh, and Dautenhahn, 2003; Fussell, Kiesler, Setlock, and Yew, 2008; Yanco and Drury, 2004). One aspect of robot socialness—and the closely tied concept of object anthropomorphization—very relevant to the current use contexts of EOD robots is examining whether or not the everyday human-robot interactions may influence operator decision-making, such as when a robot is put into a dangerous situation by the user or by the task. People are more likely to anthropomorphize robots they interact with than robots in general (Fussel, Kiesler, Setlock, and Yew, 2008). Chandler and Schwarz (2010) suggest some aspects of anthropormorphication of product design can have positive results, such as the owners' increased effort to maintain the object. However, Chandler and Schwarz also explain that in a social system people are disinclined to replace *close others*, and their findings suggest that the same averse feelings are true for replacing anthropomorphized possessions. Social norms and personal attachment are two factors that contribute to influencing this replacement hesitancy (Heider, 1958).

Furthermore, objects that hold particular emotional significance are regarded by their users as having a uniqueness that sets them apart from other similar objects, even ones that are outwardly indistinguishable from others. As such, the perceptible properties of even a mass-produced item, such as an automobile, a teddy bear, or a robot, are not the only evaluative aspects of the object that people use to process preferences. People sometimes associate properties with an object that go beyond its physical qualities, into less tangible but equally meaningful beliefs about the object, and how they use and interact with it. In these circumstances, more granular questions about attachment arise, such as which people are more likely to generate these types of attachments, and what processes of product attachment can be exacerbated or mitigated by the use case scenarios.

One school of thought about the development of humanoid robots is that there will have to be different "classes" of design, with robots appearing and acting more or less humanlike or lifelike depending on their function, role, and context of use (Clifford I. Nass, personal communication, March 2, 2006). Apart from anthropomorphication, a similar issue that triggers operator empathy toward a robot was demonstrated in an experiment using the zoomorphic robot Pleo (Rosenthal-von der Pütten, Krämer, Hoffmann, Sobieraj, and Eimler, 2012). Study participants interacted with the dinosaur robot Pleo prior to watching video of the robot either (1) tortured or (2) in a normal, untortured context. Findings showed that physiological arousal measured during both video scenarios was higher for participants who had interacted with Pleo prior to watching the videos. Research like this bears out independent reports from companies such as Sony, who produced the popular robotic dog toy Aibo that was discontinued in 2006 to the disappointment of passionate Aibo enthusiasts. Some of these toy robot owners mourned the loss not just of the robots when production halted, but of Sony's 2014 announcement of the end of its repair support for Aibo (Walker, 2015). This disintegration of the formal care system for the robot dogs not only led to the formation of Aibo-owner repair and support groups, but has contributed to the new bionic pet "veterinarian" industry.

Previous human-robot studies demonstrated that in the past, people often did not acknowledge that they saw robots as social beings when they self-reported in research settings (Kolb, 2012; Carpenter, Davis, Erwin-Stewart, and Vye, 2008; Carpenter, Eliot, and Schultheis, 2006; Nass and Moon, 2000; Reeves and Nass, 1996). Yet as shown here with Pleo, the participants also confessed to having negative feelings when the somewhat animal-like robot was tortured (2012). As robots are integrated into our everyday lives in different roles, our level of familiarity with the general idea of robots will change.

Another interesting piece of the Rosenthal-von der Pütten et al. study (2012) is that participant *loneliness* affected their level of emotions and empathy with the robot Pleo. Lonely people may employ a variety of behaviors to mitigate the pain of social isolation. Epley, Akalis, Waytz, and Cacciopo (2008) and Epley, Waytz, Akalis, and Caciappo (2008) have suggested that one way lonely people may attempt to alleviate a human-human disconnect is by anthropomorphizing non-human agents such as "mechanical devices" (p. 114). This vein of inquiry leads down a path of possible scenarios for military personnel who are separated from their homes and families during deployment and missions, and merits further study into the impact of loneliness, the humanization of familiar robots, and its impact on decision-making.

However, in the Rosenthal-von der Pütten et al. study (2012), the people who interacted with the robot prior to watching the video stimulus had ten minutes to become familiar with Pleo, an introductory period of time not conducive to forming strong bonds of attachment. One of the basic premises of Bowlby's (1973, 1980, and 1982) attachment theory is that physical or psychological threats (e.g., assessing UXO, the injury or death of a team member) automatically activates the attachment system, a system whose goal is maintenance of proximity to supportive others. Therefore, understanding when robots become regarded as "supportive others" to users is part of the key to understanding how the dynamic interactions between human and robot need to be balanced in order to create the most effective and safe HRI scenario. In addition, understanding the details of what situations, robot design, operator personality, training, and other factors contribute to the formation of the *supportive other* role also implies it will be possible to manipulate these specific factors. In turn, purposefully changing these variables leads to the opportunity to enhance or mitigate the attachment bond of human operators toward robots in a way best suited toward safely achieving missions and tasks.

Chapter 5
Robot Design as Rhetoric

Human emotions affect the way people approach and solve problems, which is clearly a function of rhetoric, and in this case, persuasive design. How people choose to design or use a product—or whether to even use it or not—partly depends on emotionally driven creative ways of solving problems. The functionality of a robot modeled on some characteristics of a human or animal can have very practical value. Yet intentionally lifelike design in robots immediately thrusts them into a role that straddles human-human/animal and human-robot relationships. *Anthropomorphism* describes the human tendency to instill the real or imagined behavior of non-human agents with humanlike characteristics, agency, intentions, or emotions, while *zoomorphism* refers to similar concepts surrounding an instinct to instill animal-like characteristics into something not necessarily an animal. It is necessary to be clear that the human tendency to attribute lifelike qualities to robots occurs when interpreting a variety of cues, and does not rely on external appearance alone.

A complex system like a robot can have many lifelike signs. Examples of inanimate object cues instilled in some robots that people may use to anthropomorphize them include robot responsiveness to (or the use of) natural language; gestures or movement interpreted as having intentionality (Breazeal and Scassellati, 1999; Norman, 2005); or having parts that resemble or work like a human body (Mori, 1970; DiSalvo, Gemperle, Forlizzi, and Kiesler, 2002). The *situation* in which a robot and human interact and the robot's role may also mimic certain aspects of human-human interaction, such as robots that assist in domestic or care settings. Nass and Moon (2000) have stated that even a minimal amount of humanlike cues can evoke a wide range of strong attitudinal and behavioral consequences in individuals. Because humans use their models of human-human interaction to make sense of robot interactions, they may overestimate robot intelligence (Lee, Kiesler, Lau, and Chiu, 2005) from minimal social signals, possibly affecting human-robot team or collaborative interactions.

How robots interact with one another may also present a variety of cues to those observing the robots about how to interact with them, or provide clues to their abilities. These relational actions provide models for our interpretation of what level of socialness is embedded in the robot, the general sophistication of the system. However, how external clues are interpreted by observable interactions and characteristics is not necessarily an accurate indicator of robots' true capabilities or limitations. Yet depending upon the types of interactions observed, people bring their own communication schemas to overlay on their observation. Robots that appear humanlike or animallike and interact with one another in ways expected from these design choices can reinforce user expectations based on what they know about humans and animals. If those robot behavior expectations are violated, it may be initially unsettling for users to process new ways of acting within these communication models because it is presenting a new and novel interaction. Furthermore, regardless of how they appear, robots that do not display social behaviors with one another may be perceived as less humanlike or lifelike.

Dunn (1995) uses the term *morphology* to describe the phenomenon of a user's perception when "the degree to which an object ... measures up to their perception of living forms, based on their own body-centric cognitive constructs about what constitutes the parts of a living form." He further explains the important concept of morphology as the assumption that people project their own meaning and experience of embodiment on to the patterns implied by the stimulus. He further states that people instill these constructs with "affection and expectation and endow them with attitudes and emotions ... . They react to, describe and remember them almost as they might other people."

Other recent studies (Carpenter et al., 2008; Carpenter, Eliot, and Schultheis, 2006) have indicated that user expectations and preferences for robot appearance correspond with Dunn's explanation, with users matching expected robot capabilities and behaviors to the outward anthropomorphic design affordances of robots used as stimulus for participant responses. The same studies indicate people also attribute agency and emotion to humanoid robots, even describing them in human terms. In other words, if a robot has something the user believes resembles humanlike hands, users expect the robot hands to function similarly to human hands. Furthermore, these design characteristics may trigger emotional associations to the robot in a way similar to that of human characteristics would. In this example, the robot hands are an affordance, and the user is matching their own mental construct of human hands and their function to that of the robot's "hands."

Robot forms, or embodiment, have been categorized into four appearance-based groups (Fong et al., 2003): (1) anthropomorphic (humanlike), (2) zoomorphic (animal-like), (3) caricatured (exaggerated qualities), and (4) functional (design based on its intended tasks). Although other social cues besides appearance impact the human tendency to anthropomorphize a robot (or not), a degree of humanlikeness in robotic form affects how people interact with robots, and can establish social expectations about interactions and abilities (Carpenter et al., 2008, 2009; Fong et al., 2003). Experimental research in human-robot collaborative team interactions has demonstrated that in these situations, not all robots are treated the same by human partners (Groom, Takayama, Ochi, and Nass, 2009; Hinds, Roberts, and Jones, 2004). For example, highly anthropomorphic robots are praised more and punished less (Bartneck, Reichenbach, and Carpenter, 2006) than other mechanical representations, such as less humanlike robots. Nevertheless, these experiments also demonstrate a need for *in situ* observations of human-robot teams conducted over extended periods to see if human attitudes and expectations change in any way toward humanoid robots used every day in collaborative situations. This type of long-term research can help determine what level of robot anthropomorphism facilitates EOD and other human-robot work, instead of impeding it.

EOD robots already imitate human behaviors just by the nature of their tasks, even when they are teleoperated field machines with low-level artificial intelligence. As critical components of the EOD tool kit, robots sometimes stand in for personnel to perform dangerous duties and help complete missions. There is anecdotal evidence that supports the idea that in some cases, EOD personnel do attribute organic traits to these machines (Barylick, 2006; Garreau, 2007; Kelly and Johnson, 2012), and describe their relationship to the robot in terms of emotional attachment, sometimes naming the robots, and treating the robot as a pet, teammate, or as an extension of themselves.

Understanding how to design a human-robot interface that will communicate naturally and effectively with a user is a significant aspect of robot design. Yet there is always a balance, a design tension, between the optimal combination of machinelike and humanlike interface attributes to support people's goals (e.g., to be assisted as needed when immobile) as well as fulfilling the robot's functional use (e.g., to be strong enough to lift a patient), including the very basic notion of communication. Rhetorical phenomena, or when a social actor explains, excuses, justifies, disclaims, or neutralizes their actions, are open to evaluation by oneself or another in any interaction. These phenomena are intersubjective because they operate around the awareness of other social actors who have the capacity to evaluate the interpretation and thereby make the social actor accountable and visible within a particular moral framework. The ambiguity of communication and interpretation in these interactions can work for or against the likelihood of a cooperative task being executed efficiently and effectively. Subsequently, while an only vaguely humanlike face on a robot may elicit positive social circumstances in one use case scenario, if that "face" also defies user expectations it can impede the interaction as well. Therefore, a roboticist thinking about the very nature of designing a robot to communicate effectively must consider rhetorical phenomena throughout development.

It is also important to clarify the word *design* as it is used here. Rosenman and Gero offer a relevant definition: "Design is a purposeful human activity in which cognitive processes are used to transform human needs and intent into an embodied agent ... Design is about the transition of concepts from the sociocultural environment to the description of technical objects." This definition combines its direct correlation to embodied agents with a sense of sociocultural awareness.

Humans have historically tended toward either extremely positive or extremely negative attitudes toward novel communication technologies (Feigenbaum and McCorduck, 1982), at least initially. Although industrial robots like those used in factory settings have been around for a period in recent history, robots interacting with people in their everyday lives are still a new medium of communication technology. Incorporating lifelike dimensions into a robot's design adds complicated layers of meaning to user expectations, and it is therefore reasonable to claim that there is the possibility people can have negative interactions with a robot based on their expectations, if the robot does not meet the users' anticipated appraisal of performance or behavior, even for utilitarian or repetitive tasks.

Social acceptance of lifelike robots is an extremely complex prospect. For a robot to engage efficiently in human-robot cooperative situations, it is required to possess a certain degree of humanlike qualities in its design, whether in form or behaviors. Responding to a natural language command increases the utility of a robot, and opens its use to others not specifically trained in the mechanics of other forms of robot operation. That same ability also can save time for an expert operator, eliminating the need for hands-on control. However, while incorporating any lifelike characteristic in a medium that moves in our space and appears responsive to our needs simultaneously offers signals to users about how to interact with it in an efficient way, it also represents other familiar characteristics as associations, particularly of living things.

Determining what degree to which these lifelike qualities should be purposefully imbued into robot design is not an easy problem to resolve. The introduction of these social cues increases the likelihood of user expectations exceeding the system's actual performance and abilities, at least in initial interactions. It is indeed possible, as Mori hypothesized, that making androids too humanlike will repulse some users. In addition, highly humanoid robots can give an impression of intelligence so superior to the user that they become undesirable objects to interact with, a resistance from the user that may be difficult to overcome quickly. From some designer perspectives, there is the desire to mimic life as a form of creative outlet as well. Roboticist David Hanson remarked, "Nobody complains that Bernini's sculptures are too darn real, right? Or that Norman Rockwell's paintings are too creepy. Well, robots can seem real and be loved, too. We're trying to make a new art medium out of robotics" (Slagle, 2007).

Recognizable aspects of humanlike appearance and social characteristics incorporated into products and tools encourage us to see these things as more than simply objects, but also as believable, appealing social actors. Thus, one of the keys to the successful design of social humanoid robots meant to be used in a field defense setting is to find the balance in design that leads the user to believe in the abilities and sophistication of the system and trust in its tool characteristics, while not eliciting unintended expectations about the robot's actual functionality.

Yet in most situations, it will not be desirable to design robots without social intelligence when they interact with people in cooperative scenarios. Omitting all socialness from the interactions will add to the user's cognitive load, requiring them to operate and interact with the robot in a completely different manner than with human team members. The addition of some socialness eases the human side of the co-participation process.

Still another reason to add social capabilities in robot design is that from the machine perspective, learning can be simplified. Programming robots is challenging, and if robots have the ability to learn via imitation, experience, feedback, or other social processes, it relieves some of the need to anticipate and program for every social cue or task. A human operator will be able to correct the robots' actions on the fly if necessary, until the robot learns through situational understanding how and when to use certain strategies and behaviors.

Similar to humans, for robots, learning is the internalization and application of skills, tasks, and information. However, humans perceive the world through very different sensory modes, and so use human-centered insights about what is significant to know or understand. Knowledge is framed by experience in the world, as well as values and a particular set of expert skills. This information is carried and applied adroitly on to what is already known about different contexts. Prior experience is used and the contextual information melded together to build a framework of evaluation about what to attend to, what to do, what order to do things in, and to predict what is expected will happen as a result of any actions.

Social aspects of learning go beyond mimicry or imitation. The social characteristics of human pedagogy can also be an effective method to build robot intelligence, if the robot can learn from fewer examples and then generalize strategies to other situations. A robot's ability to accurately generalize its experience means it can then apply learned strategies to other situations with adaptations without specific direction from an operator. This kind of learning intelligence also means a robot can learn to fix problems in a variety of situations, and predict inaccuracies in similar systems before they occur.

The aesthetics of robot appearance, its behaviors, intended social role, methods of communication, functions, and context of use all have an influence on user engagement. Engagement is closely tied to believability. The user's level of involvement or potential for finding the android appealing is one definition of engagement, and it indicates how much the user is attracted by the android or how familiar or distant she feels to it.

Designing one perfect, off-the-shelf model of a robot for use in any scenario that is successfully engaging to a wide variety of users is not likely. As with many complex technologies, options for easy modifications and personalization will increase not only the potential uses for a robot, but also create additional opportunities for user engagement. In addition to the modification of capabilities like armor or appliqué, the ability to change characteristics like robot voice, gaze, or aspects of its appearance will influence what is sometimes referred to as its *visceral design* (Norman, 2004). Here is where it is especially useful to have a design vocabulary that describes different parts of a user's whole perception of a product, since this is a fluid process and not easily parsed out. As part of the biology of human affective processing, at a visceral level people make rapid judgments about what in the environment is safe or threatening, sending signals to the muscles and the brain. At a behavioral level, without conscious choice, reflective design is used thought to guide actions. Furthermore, *behavioral design* refers to the user's gratification when interacting with the product, while *reflective design* is tied to a user's own self-image and memories related to the object. Using this model of understanding human-product attachment, behavioral design is related directly to a product's use and its performance, while reflective design is a connection to the user's self-image.

Robot morphology refers to more than the external shape. Here, morphology describes user perceptions of objects, and whether these objects match up to user expectations of a living form, due to the human-centered way of using cognitive constructs of what has been experienced in order to scaffold an understanding of is not immediately comprehended (Dunn, 1995). In the case of determining whether a robot is believable, or something that will function as expected, people turn to a model of what they believe they know best: the human form or humanlike way of interacting with the environment and the world. *Believability* is then not necessarily the idea that a robot appears to be lifelike, but more accurately describes the concept that the robot matches the users' expectations of robot shape, and therefore its abilities and qualities. There is an innate process of projecting meanings and experience and sense of embodiment on to the model provided by a stimulus—in this case, a robot. With believability, associations often carry other facets that the construct has in place, including affection, and even applies emotions and attitudes to the socially believable interactive product.

Our morphological perceptions can be further divided into groups. *Physical form* refers to perceptions based on facets of the object's appearance, such as structure, symmetry, distinct body shape, monadic structure, proportions, and whether it is a single thing. If a thing appears to have a sensory system, such as eyes, a mouth, or even simply something resembling a head or face, it is part of the user's morphological perceptions of the other thing. As with attachment to people and things, our perceptions about an object's *dynamic form* evolves over time, and includes characteristics like coordination, quality of motion, verisimilitude, and scale.

To balance a robot's physical feature designs so its effectiveness is preserved while simultaneously designing in such a way that it does not violate user expectations is an enormous challenge. Again, one method used to guide robot development is using successful human-human models of interaction, in this case applying components such as visibility of physical cues, actions, and internal states, awareness, and accountability. A concept like accountability is further embedded within the rules, norms, customs, and basic social mechanisms for control. These characteristics of humanlike behavior let users facilely draw from their own social experience and structure their interaction with a robot, similar to a familiar human-human interaction. These aspects of robot socialness are pieces of visible information that also may encourage the user's level of engagement with the robot. Similarly, the designer must avoid representing "fake" emotions. Imbuing a robot with emotions that seem insincere conflict can conflict with a user's idea of how an interaction should happen within the context of use, possibly implying a sense of the robot's functional abilities, or even eliciting a feeling of deception. Either of these negative associations can impede the user's trust in the robot, or their interactions in general.

Perhaps not surprisingly, because it seems so instinctual and innate a thing, humans are naturally attuned to faces. There is a reaction to the play of emotions across a face, an expectation of clues that accurately reveal an inner state. In this sense, the use of humanlike cues can facilitate the social understanding of another.

The design features that imply things like a head, or eyes, or a mouth, are also additional affordances, or cues that an interaction with this object will be based on a natural or organic way of communicating. Similarly, conversational turn taking will require a robot to consistently and accurately assess human voice tones and other vocal behaviors, in order for it to apply. The nuances of these physical attributes and behaviors also closely relate to the idea of *personality*, whether this is an intended perception from the design viewpoint or not.

The term *invisible machinery* has been used to refer to robots that, through humanlike external design and behavior, produce a sense in the user that the object is natural or living (Carpenter et al., 2008). Unlike Mori's theory of the uncanny valley, invisible machinery does not necessarily attach a sense of discomfort about conflicting human perceptions of naturalness and artificialness embodied in the same entity, but simply identifies the phenomenon, and in a way specific to mechanical systems, particularly robots.

Similarly, the Cockney argot idiom *clockwork orange* refers to a person who acts strangely or different, although they appear to be like others. Fiction author Burgess (1986) explained that his literary allusion to a clockwork orange was meant to "stand for the application of a mechanistic morality to a living organism oozing with juice and sweetness" (p. xv). Moreover, Burgess explained his use of the phrase in terms of moral choice: "If he can only perform good or only perform evil, then he is a clockwork orange—meaning that he has the appearance of an organism lovely with colour and juice but is in fact only a clockwork toy to be wound up by God or the Devil, or (since this is increasingly replacing both) the Almighty State" (p. xiii).

Both of these explanations of a clockwork orange have parallels to the premise of invisible machinery. Invisible machinery is, in some ways, also a human-centered concept when it is applied to perceptions of robots because it suggests how people sometimes experience something artificial as something natural. Whether or not robots are imbued with or develop a unique and new version of morality may be on the horizon, since moral choice is at the center of many decisions, small and large. Like a clockwork anything, the process of robot creation is socially cooperative, and undertaken by very human hands. From the roboticists' perspective, intentionally or not, creating invisible machinery is akin to transforming a machine into something people will recognize as *more* in the eyes of the beholders. Yet, from Burgess's example, the audience views the clockwork orange as a disappointment, something that has been inaccurately interpreted because it resembles something natural when it is not. At this point, it is unknown how people will view invisible machinery over time.

While the clockwork orange may also be seen as a symbol for the world, invisible machinery refers specifically to how human-robot interaction perceptions may inform how the world is viewed. Another distinction between the two expressions is this: a clockwork orange is a metaphor, and no such fruit thing really exists. However, it is likely robots perceived as invisible machinery will soon be part of our everyday world.

PART III
Patterns

Things that are learned within different social systems influence emotions, assessments of context and situations, and behaviors. Therefore, when designing the most effective robots, there should be research examining as much of the social systems the users are embedded in as possible. Social systems are a group of people in a collective whole and their related group roles, and describing these groups does not necessarily focus on the individual members. Within these social groups are agreed-upon formal and informal values and norms that establish agreed-upon social guidelines within the group context. In order to analyze what is happening in one social system, there must be a close examination of the existing patterns or relationships connecting these parts. A major goal of this type of research is to gain an understanding of the everyday contexts, individual and group experiences, and interactions that currently take place by describing how these factors ultimately shape human-robot interactions at micro levels.

The nature of Explosive Ordnance Disposal (EOD) work is distinctive within military specializations. EOD personnel go through some of the most rigorous specialized training in the military. The initial training period includes a component all members of the Armed Forces attend together in the Naval School Explosive Ordnance Disposal (NAVSCOLEOD) located at Eglin, Air Force Base (AFB), colloquially referred to as "The Schoolhouse." Once graduated into the formal EOD role, the job demands academic and physical prowess.

Like in some other military specialties, small-group teamwork is critical to the job, but EOD work also demands ongoing effective verbal communication between group members in order to successfully complete the team-oriented missions. Also unusual in the context of more typical military structure, EOD team members are frequently encouraged to give input to the Team Leader about each mission's situation while it is in progress. This procedural approach is based on the assumption that every individual is a Subject Matter Expert (SME) with a valid perspective worth considering in a collaborative effort before the Team Leader decides on the final group actions.

As EOD training and work evolves within the military, and because of a surge in Improvised Explosive Device (IED) encounters, modifications are being made to aspects of EOD teamwork. These include team size, the age of people promoted, and increased reliance on technology such as robots. One of the most critical standard tools EOD personnel use are the semiautonomous teleoperated robots that assist in Render Safe Procedures (RSP), helping to disable or mitigate the threat of explosives.

Currently, EOD personnel increasingly use robots as an important tool to assist in render safe procedures for unexploded ordnance. Consequently, if problems with the human-robot interactions are overlooked, there is a continued danger to human lives and mission outcomes from the unidentified issues in these interactions.

The EOD system is one of the dynamically interdependent pieces of the holistic experience viewed as an *ecological system* (Bateson, 1972; Sundstrom and Altman, 1989; Sundstrom, De Meuse, and Futrell, 1990; Murphy, 2004; von Bertalanffy, 1968) of interconnected parts. *Patterns* presents background and context of EOD work via description of the organization, people, robots, operating environment, and tasks as they relate specifically to EOD work, in order to present a rich description of impactful parts of U.S. military EOD systems. Furthermore, for qualitative research data such as that gathered from the interviews conducted with EOD study participants, the information must be grouped into meaningful patterns or themes that were observed. This iterative process is the core of thematic qualitative data analysis. Following this framework, the world of EOD personnel is explained in terms of how they are situated within the military as an organization; characteristics and shared experiences of EOD as a group; how robots are typically used in EOD work; the nature of an EOD operating environment; and the way EOD manage cooperative, behavioral, and conceptual tasks.

EOD human-robot work presents unique emotional challenges that must be considered as robot design and team size evolves. These interactions include how emotion in human-robot interaction affects operator decision-making and, among other things, mission outcomes, which are life-and-death situations. In the National Research Council's report (2002), a call was made for continued robotic development for ordnance disposal, as well as an increased focus on the human factors side: "Because technologies are implemented and operated by human agents and social organizations, their design and deployment must take human, social, and organizational factors into account" (p. 298). Much has been written on the topics of military cohesion, inter- and intragroup relations, and the engineering side of robot development, but relatively little has been published about the nature of EOD work, how robots are integrated into their world, and how troops work with robots. Therefore, in order to begin to understand individuals working within Explosive Ordnance Disposal, it is critical to examine the various interconnected parts of the system in order to provide a framework for inquiry at a theoretical level.

In the United States military, Explosive Ordnance Disposal (EOD) technicians perform a vital role, effectively and safely defusing U.S. and foreign chemical, biological, radiological, and nuclear (CBRN) unexploded ordnance (UXO), including Improvised Explosive Devices (Department of Defense, 2006). U.S. military EOD specialists also work stateside assisting local and state civil authorities to disarm and dispose of hazardous devices. Various other official EOD responsibilities include support of the U.S. Secret Service, State Department, and other Federal agencies (Cooper, 2011; United States Army, 1997). These include the U.S. Department of Homeland Security, U.S. Customs Office, and Bureau of

Alcohol, Tobacco, Firearms, and Explosives (ATF). Their services protect the president, vice president, and other officials and dignitaries, as well as providing a critical security role at large international events.

United States Armed Forces Explosive Ordnance Disposal specialists also train and assist domestic civilian law enforcement personnel (Larry, 2008, para. 2; National EOD Association, 2012; United States Army, 1997) and international friendly and allied force military EOD specialists (Gibson, 2009; Owolabi, 2010, para. 1; Valentin, 2011).

EOD personnel are relatively new to the U.S. military, but have found an unfortunate new significance in recent years due to the increased use of Improvised Explosive Devices in warfare. Improvised Explosive Devices are essentially homemade bombs, often positioned roadside in a very grassroots, non-military fashion. IEDs are an alternative style to conventional weapons, often built by informally trained people, and ironically proving to be as dangerous—or more—than standardized military tactics. Individuals and groups using IEDs as a preferred method of provocation and destruction may adapt their low-grade technology quickly, accompanied with the associated tactics, techniques, and procedures (TTPs) evolving in shorter and shorter cycles (Wilson, 2007). IEDs vary in design and may contain many variations of components, such as detonators and explosive loads. Typically, antipersonnel IEDs include shrapnel-generating objects such as nails, or disburse harmful chemical agents.

Mines, in contrast to IEDs, are usually based on a conventional design and are standardized and mass replicated. IEDs take many forms, and are triggered by an assortment of methods, including infrared or magnetic triggers, remote control, or pressure-sensitive bars or trip wires. Multiple IEDs are sometimes wired together in a daisy chain to attack a convoy of vehicles along a road. There is always the threat that toxic chemical, biological, or radioactive material may be an added component of the explosion, creating other severe effects beyond the shrapnel, concussive blasts, and fire normally associated with bombs.

Variations of IEDs include the Vehicle Borne IED (VBIED), commonly known as a car or truck bomb, and the House Borne IED (HBIED), created when an entire home or similar structure is rigged to detonate. Insurgents will often watch any EOD investigative activities in order to set off the explosive strategically and detonate it remotely to cause the most harm, or use the IED to lure EOD personnel into the range of sniper fire. Lieutenant General Michael Barbero, Director of theJoint IED Defeat Organization, has said, "In the 20th century, artillery was the greatest producer of troop casualties. The IED is the artillery of the 21st century" (JIEDDO Counter-IED Strategic Plan, 2012-6).

A critical tool in the first line of any EOD teams' defense against these threats is the use of various mobile robots that perform dangerous tasks such as Unexploded Ordnance (UXO) disposal, vehicle inspection for hidden IEDs, and advance scouting of dangerous transportation routes. The key reason to use robots to detect, inspect, or disarm IEDs is to distance EOD personnel from the danger, thereby reducing the chance of human injury or death.

The rise of IED use cannot be overstated. In 2007, IEDs caused over 70 percent of all American combat casualties in Iraq and 50 percent of combat casualties in Afghanistan, including fatalities and wounded (Wilson, 2007). The impact of IEDs on civilians in areas of conflict has increased alarmingly in recent years, too. The United Nations Mission Assistance in Afghanistan (UNAMA) officially stated, "In incidents where intended targets appeared to be military, those responsible for placing or detonating IEDs showed no regard for the presence of civilians and no evidence of distinguishing between civilian and military targets in violation of the international humanitarian law principles of distinction, precaution and proportionality" (2012, p. 10).

UNAMA reports that IEDs are the biggest cause of death in Afghanistan's armed conflict, and recorded the deaths of 340 civilians as well as 599 additional injuries from January to September of 2012 (UNAMA press release). The *Afghanistan Annual Report on Protection of Civilians in Armed Conflict* stated that in 2010, 40 percent of female civilian deaths and 44 percent of child deaths were a result of IED explosions and related suicide attacks (2011).

According to an official of the Department of Defense's Joint Improvised Explosive Device Defeat Organization (JIEDDO), military data reports insurgents in Afghanistan plant up to 1,400 IEDs per month (Dreazen, 2011). At the peak of the Iraq war, there were over 4,000 IEDs planted per month (Mora, 2010). Outside of Iraq and Afghanistan in 2011, from January to November, there were 6,832 IED events globally, averaging 621 per month, resulting in 12,286 casualties in 111 countries (iCasualties.org, 2011; JIEDDO, 2012a). Excluding Afghanistan and Iraq statistics, global IED casualties reached their peak in May 2012 with approximately 1800 people wounded and almost 600 killed in May alone (JIEDDO, 2012b).

In 2010, the U.S. military increased the number of road-clearing teams in Iraq from about 23 to 56 (Flaherty, 2010) and in 2011, augmented the road-clearing troops in Afghanistan from 12 to 75 (Dreazen, 2011). Reported numbers vary, but there are currently about 3,000 tactical robots in Iraq and Afghanistan used for reconnaissance and UXO sweeping (Osborn, 2010, para. 1; Singer, 2010), with about 2,000 ground robots in Afghanistan alone (see more detail on pp. 9–10), a ratio of approximately one for every 50 troops (Axe, 2011).

Unfortunately, these statistics tell only part of the story, as IED threats increase as a method of warfare and terrorism every year, and represent greater-than-ever hazards to worldwide military personnel, domestic first responders, and civilians. Although EOD personnel work in many different situations with a comprehensive range of ordnance, the regrettable increasing popularity of IEDs as a weapon has been a significant springboard for some of the rapid changes within the EOD field in terms of recruitment, training, team structure, and tools used.

The Ecological System of
U.S. Military EOD Work

Organization

This work delineates an *organization* as a definable group of people with a shared history that has a culture, as well as collective values and norms (Rousseau and Cooke, 1988). It is important to distinguish *culture*, or the pattern of meanings embedded in symbols, from *social culture*, which is the "economic, political, and social relations among individuals and groups" (Geertz, 1973, p. 362), although both cultural aspects of EOD life are examined here in a broad scope. If an organization as a whole has had shared experiences, a total organizational culture will exist. Similarly, if an organization has subgroups with shared experiences, many subcultures can arise.

The military consists of subgroups, units, and teams in many forms (Arrow, 2000). In lieu of attempting to scrutinize every value, symbol, artifact, and assumption made by the larger group and all subgroups in this section, *culture* is used here to explore how EOD personnel learn appropriate individual actions within their inter- and intragroup experiences through the shared history of formal and informal training, doctrine, rituals, and practice. Therefore, this broad concept of culture is described and analyzed by examining some examples relevant to these identified aspects of culture at the organizational level and in terms of its impact on EOD personnel.

Each service branch's actual description of their own EOD forces varies little, and shares a common mission that encompasses the protection of personnel, facilities, and critical infrastructure from the hazards posed by Unexploded Ordnance (UXO) and IEDs during combat operations, in peacetime, and in foreign and domestic settings (Department of Defense, 2006). The term *UXO* also refers to U.S. and foreign chemical, biological, radiological, and nuclear (CBRN) ordnance.

Under this broad scope of potential circumstances and situations, the settings and tasks for EOD specialists vary according to mission. These technicians tactically assist other military personnel by reducing UXO and IED threats from principal lines of communication and supply routes. They are the only forces specifically trained, equipped, and tasked to remove or lessen the hazards posed by UXO and IED and to also train non-EOD personnel how to recognize potential UXO threats. They are frequently called upon to assist specialist EOD police units; they dispose of old or unstable explosives, such as ones used in mining, fireworks, and ammunition. As highly trained ordnance experts, they are responsible for escorting diplomats and dignitaries, sweeping VIP-traveled areas for UXO, and ensuring the safety of other public places during large events. Another task of EOD technicians is to conduct post-blast investigations. The EOD job responsibilities also include supporting government intelligence units and federal agencies such as the U.S. Secret Service and the State Department.

In their effort to reduce the threat of UXO and IEDs, the use of specific EOD tools and methods to prevent detonation—or the Render Safe Procedure (RSP)—is the critical step toward the goal of creating a secure environment for military personnel and civilians. RSP is the set of actions to render safe unexploded ordnance or an improvised explosive device based on the training, experience, and situation or mission of the EOD technicians and their team, using specific technical procedures, tools, and methods (Air Land Sea Application Center, 2001).

While the threat of IEDs has been growing exponentially in recent years, the number of troops who are specially trained to defeat this threat has not grown at the same rate, although there have been increased efforts within the military branches to raise awareness of EOD work as an option (D. Brown, 2000; Svan, 2008; Talton, 2008). As of 2008, there were 456 U.S. Marines with the primary military occupational specialty of EOD technician, although the stated goal was 663 with completed EOD training (Svan, 2008). Svan (2008) also reported that the rate of attrition had sometimes outpaced the rate of those entering EOD training, and the U.S. Navy's 912 EOD enlisted numbers were at 86 percent, or about 152 people short of their desired numbers. Since Svan's report, there has been an increase in Navy EOD personnel to 485 officers and 1105 enlisted sailors, according to an official Navy post (Explosive Ordnance Disposal Group 1, 2013). There are currently approximately 1,800 soldiers in the Army's Explosive Ordnance Disposal units (Hall, 2011).

One way to establish common ground is to provide universal training (Mark, 1997) and formal education. The formal education that EOD personnel take part in includes classroom learning with an instructor and an established body of curriculum, as well as hands-on activities. However, EOD personnel continue learning activities throughout their career, including peer-to-peer training, self-directed activities, and formal military courses that are required or optional. They also receive additional training to keep their knowledge current, or for advancement and further specialty certification. In this work, *formal learning* and *formal training* refers to schooling activities that reward with formal credentials, and are taught by an official military-appointed instructor. *Informal leaning* and *informal training* occur outside the scope of the formal requirements, and are, as Livingstone (2001, p. 4) explained the term, "any activity involving the pursuit of understanding, knowledge or skill which occurs without the presence of externally imposed curricular criteria."

In other words, the basic curriculum, goals, outcomes, and method of knowledge acquisition used in informal training and learning may be determined by an individual or collectively by a group of people, and is often self-directed to some degree. However, the basic difference between formal and informal

Chapter 6

learning as it is used here is that informal learning and training take place without an institutionally recognized and appointed instructor. Throughout this book, the term *informal learning* refers to both self-directed informal learning and informal education/training unless otherwise specified.

In the United States military, Explosive Ordnance Disposal personnel undergo some of the most comprehensive qualification training in all of the branches' occupational specialties. All military personnel go through some form of induction training to acquire fundamental skills and receive a certain level of indoctrination by learning service norms, procedures, specialized language, and symbols (Vygotsky, 1986). This training period formally begins with Basic Combat Training (BCT), otherwise referred to as Recruit Training or colloquially known as Basic (U.S. Army, *Soldier Life*, n.d.). This initial training lays a foundation for the individuals' assimilation process into the military by de-emphasizing the person as a solitary unit and emphasizing group work, progress, and shared goals (Janowitz, 1972), as well as introducing a common vocabulary and set of procedures for doing everyday things (e.g., folding clothes) and new things (e.g., operating semi-automatic rifles).

At this point, a new recruit's success is not based on prior academic achievement or socio-economic status. Training may occasionally take the form of coaching and focus on an individual that needs assistance with a specific skill, such as the physical requirements, but frequently a form of in-group apprenticeship is used to bootstrap the person's ability to achieve. Therefore, comradeship is built among people of diverse backgrounds and the person becomes part of a greater whole.

After choosing to commit to EOD work, on average, in all service branches of the U.S. military, EOD technicians spend 10 months in job training at various locations depending on requirements that are branch-specific, such as the Navy dive school and parachute jump training. Navy EOD specialists participate in an approximately year-long odyssey with weeks of academic and physical preparation work, Basic EOD Diver Training, 42 weeks of Basic EOD Training, three weeks of EOD Tactical Training, and three weeks of U.S. Army Jump School.

However, all four branches at some point funnel through basic EOD training at Eglin Air Force Base (AFB), Florida (Cooper, 2011). In 1999, the Naval School Explosive Ordnance Disposal (NAVSCOLEOD)—Navy-managed command staffed by members of all military branches—centralized the basic EOD training at Eglin AFB. "The Schoolhouse," as EOD personnel sometimes refer to the school at Eglin, trains about 800 students each year, according to the official Web site (Naval Explosive Ordnance School, 2011, para. 5). At Fort A.P. Hill in Virginia, there is a new training ground for the Army and other branches of the military: a 2,700-acre Explosive Ordnance Disposal range (Dennen, 2011), further expanding EOD resources while centralizing additional training options.

It is interesting to note that currently EOD students work hands-on with robots during only a small part of the official EOD student training program. Recently, virtual video-game based platforms have been incorporated into training (Robillard, 2011), using the same joystick and controls that a fielded robot has, but via a virtual environment to facilitate learning how to assemble the robot and maneuver it in different situations.

In all arms of the military services, EOD training continues throughout their careers. Personnel may spend their downtime practicing with robots in homemade obstacle courses, reading and learning from incidence reports, and participating in Team Leader-initiated exercises. In addition, they take part in specialized training courses like Mission Rehearsal Exercises (MRX), Global Anti-Terrorism Operational Readiness (GATOR), and Team Leader certification training (Riemer, 2008; Bailey, 2011). Formal continuing education may also include advanced foreign language studies, advanced individual weapons training, combat life-saving medical skills, intrateam communications capabilities, and training specific to urban or other environments.

Even with an EOD school attrition rate of approximately 50 percent across service branches (Lamance, 2010; Cooper, 2011) the visibility of EOD work as a potential career track has increased due to a rising public awareness in popular culture through movies such as *The Hurt Locker* (Bigelow, 2008; Cooper, 2011; Vowell, 2013) and almost daily worldwide news reports of IED incidences. Because of the surge in international IED activity and the need for a steady supply of EOD personnel despite the high attrition rate, the very structure of training and even EOD teams is evolving.

In order to combat the high attrition rate of EOD school completion, in 2011 the Air Force initiated a 20-day "screening course" (Kelsey, 2011) to prepare students before getting to Eglin. Another strategy used to supply demand for qualified EOD personnel is experimenting with different-sized teams and lowering the required age for Army troops entering EOD training (Spencer, 2011). As Spencer (2011) points out in her article, these organizational strategies have received critical reviews from some EOD personnel. For example, it is possible that larger teams will be an impediment to bonding and communication, although there is a potential advantage of more individual perspectives in such dangerous tasks as EOD work. Similarly, Spencer points out criticisms of the lowered age requirements, most notably in the Army, where qualified 18-year-olds can now begin EOD training.

In 2002, Lt. Col. John Stefanovich warned that qualified EOD personnel are not an easily created commodity:

> Three lessons are evident from the observations made of EOD operations in Afghanistan: A well-trained EOD soldier is more important to success than any EOD equipment. The quality of an EOD soldier is more important than the quantity of EOD soldiers. Competent EOD forces cannot be mass-produced after a crisis occurs.

Although younger team members may be eager to sign up for schooling, it is too early to see if the lowered age significantly affects attrition rate or long-term overall team dynamics.

The relatively low number of qualified EOD technicians within the military branches has helped to create a very strong subculture of specialization, experience, cohesion, and in-group solidarity (Blankenship, 2011; Kirke, 2009; Yarbrough, 2008). As stated previously, these specialists have the unique position within the military where they are required to work and communicate as teams for decision-making, as opposed to the common military strict hierarchical structure where team members may have little input on a leader's chosen course of action.

EOD is also one of the few jobs where the senior members take some of the more physical risks than junior team members, and this dynamic allows the trainees to learn from the Team Leader's experience. EOD personnel are encouraged to share information through ongoing team communication (Department of Defense, 2006), where the dynamic intergroup communication critically affects the decision-making processes and therefore, mission outcomes. Written communications and reports are also used to gather intelligence, document mission findings for analysis, and inform future training scenarios (Department of Defense, 2006).

It is important to consider the significant function of a sense of community among EOD members when considering the possible human emotional connection to EOD robots as the human-robot interactions evolve and robot designs change. EOD specialists can form interpersonal and intragroup bonds through common training, shared experiences, and strong organizational ties.

Meaningful rites, rituals, customs, and traditions that are specific to the organization are also an intrinsic set of individual unifiers and social signifiers, used as powerful ways of communicating and reinforcing organizational cultural assumptions. Yet the meaning of the rites and rituals can reflect the views of a limited part of any organization, or vary in the degree to which their significance resonates as truthful, believable, or useful. Rites, rituals, and traditions follow a *social reality* that reflects some level of agreed-upon consensus about truth. Part of these ceremonies, customs, and similar behaviors is the history that accompanies them, often in the form of a type of origin story about people or experiences relevant to the organization. As the organization develops, the historical narratives evolve, and these stories, like their related rituals, reinforce cultural assumptions both within and outside an organization.

The Commander's Message of a recent issue of the *Newsletter of the National EOD Association* (Jiminez, 2011) is a formalized example acknowledging and describing the close interpersonal associations created via EOD shared experiences:

> Our members are our strength. They are the ones who joined together to form our Association. They are the ones who served in EOD throughout the world to make it a safer place for others and to enable those engaged in combat operations. They are the ones who provide the stories and pictures we cherish. They are the ones who gather from time to time with a feeling that they have never been apart—though years may have passed. They are the ones who, although they have never met before, feel a common bond when they meet. A bond forged in shared training, experiences, and danger. As our older members

pass on—we honor them. To our current members—we offer our support and our thanks for their willingness to share in the work of the Association. To our new members, we offer a hearty welcome and our hope that their membership will be a rewarding experience. (p. 1)

This organization newsletter introduction demonstrates the officer's way of building vertical cohesion within the EOD group by promoting peer bonding via supervisor support (Siebold, 2007). The statement is also gracefully inclusive of new members to the group, as well as veterans.

The "EOD Prayer" (Schott, 2011) written by Reverend Carl Bergstrom and commonly included in official EOD professional websites such as the National EOD Association (http://www.nateoda.org/) includes the phrase, "Grant that in the EOD Family there may be unity of spirit for the well-being of all." Regardless of individual religious affiliation, the sentiment appeals to the ethos and pathos of EOD work as a shared, family-type experience with the common goal of safe and healthy welfare for everyone in the circle of EOD work.

There are also a number of observable concrete examples, or artifacts, of collaborative EOD social exchanges that are actively intended to foster intergroup relationships (Gergen, 1985). Continued community building across geographic locations and over time is facilitated by the numerous online EOD personnel social groups mediated through sites like Facebook, LinkedIn, personal blogs, and bulletin boards. Sometimes the groups are established as the online face of specific battalions or associated with a military branch official Website, such as the 3d Explosive Ordnance Disposal (EOD) Battalion (Bn) Facebook page. (n.d.; https://www.facebook.com/pages/3d-Explosive-Ordnance-Disposal-EOD-Battalion/183756438317677), which is administered by 3d EOD Bn staff.

Typical posts to battalion-, platoon-, or company-specific Facebook sites include photos of graduation ceremonies from EOD training and pictures of deployed members, family-oriented event announcements for relatives and friends of serving EOD members, and platoon or company reunion announcements. In addition, notices of injury and death of EOD personnel, current and retired, are frequently written and posted in a newspaper-like obituary style with details of the individual's military career, personal life, and cause of death. Individuals have also set up informal online social spaces not directly affiliated with the military, and they have wider membership parameters. One example is LinkedIn's professional networking groups that allow, with moderator approval, any personnel with EOD experience to join, reconnect, and discuss military and civilian career opportunities around the world.

Another facet of EOD culture that cannot be overlooked because of its significance in building community among military EOD personnel is the EOD Memorial Foundation. The current iteration of the main physical EOD Memorial is in Eglin AFB, across from the main EOD building. All technicians or officer graduates of an approved EOD School who have died on active duty related to an EOD mission or duties since World War II are eligible to have their name included on this memorial wall (EOD Memorial Foundation, n.d., 2009, n.d.). Subgroups of branches and companies sometimes have their own memorial walls, including very personal artifacts like ID tags (Choate, 2011), photos, and biographies.

These memorials are meaningful reminders for the EOD groups that aid the understanding of group history by keeping the fallen group members significant to the living, and evoking appreciation and understanding for the personal sacrifices of the subgroups (e.g., Marine EOD) and groups (e.g., EOD personnel, military, American). Master Gunnery Sgt. Michael C. Sharp indicated the meaning of the wall in very plain language: "I believe the wall is a statement that the EOD community is small and the members of that community will always do our best to represent and honor our brothers and sisters in the field" (Choate, 2011).

The online presence of the Foundation contains EOD news, obituaries, and a storefront for merchandise. In addition, a database of all fallen EOD personnel of the four military branches is accessible. The linked names of the deceased reveal poignant messages from friends and loved ones. The site also hosts information about the EOD Memorial Scholarship Fund, explaining, "Applicants must be the child, stepchild, spouse, or grandchild, or other DoD recognized dependent of a graduate of NAVSCOLEOD" (EOD Memorial Foundation, 2011, para. 11). Again, this is a formal demonstration of inclusion of EOD personnel families into the larger group membership.

People

Individuals within the EOD community must have the ability to work as part of a team, have effective interpersonal communication skills, be able to withstand prolonged academic demands and physical rigor, make quick decisions based on Render Safe Procedures, and, on occasion, act *in situ* with little outside-group guidance. In other words, the people within the EOD groups operate effectively under stress and are experts on a range of topics, from engineering-oriented matters to inter- and intrapersonal skills and physical prowess. Understanding what type of person succeeds at such a multidimensional and demanding job will give richer insights into determining efficient task-oriented robot design, especially when designing a semiautonomous robot that functions as an effective tool for team collaboration without user distraction or impediment to goals.

In a 1985 longitudinal study, Hogan and Hogan (1989) identified consistent personality characteristics among Navy divers as well as Army and Navy EOD apprentices and students who successfully completed training, including traits such as well-adjustedness, spontanaeity, physical self-confidence, openness to new experiences, rowdiness, technical orientation, and introversion. However, it appeared once trainees were incorporated into the fleet, they became increasingly cautious and conforming when compared to the study's EOD students. Given these changes in personality once situations and experiences changed from classroom to fleet life, it is a reasonable hypothesis that these traits, although consistent, also evolve as the EOD personnel's knowledge changes over time from novice to expert.

More recent studies about EOD trainees and personnel produced statistically significant findings that supported the hypotheses that people attracted to and successful in EOD work have similar learning style preferences and intelligences (Bates, 2002; Bundy and Sims, 2007). Yet data also suggests that variations exist both within and between an assortment of demographics such as age, rank, or armed forces branch. Again, additional research still needs to continue in the field with EOD personnel in order to account for situational conditions and stressors beyond the classroom or training context.

Like any profession, EOD work as a whole is composed of a collection of individuals with a wide variety of ages, formal education, career goals, personal preferences, personalities, physical differences, and levels of technology acceptance. In this work, it is relevant to discuss emotions as a means to discover patterns from the group in personality type, artifact interpretation, situations, and specific goals as a way to predict or manipulate emotion through robot design (Lazarus, 1993; Norman, 1988; Rafaeli and Vilnai-Yavetz, 2004) and not as an attempt to categorize the illogical or irrational. Studying the relationship between emotion and decision-making specifically in EOD human-robot scenarios also makes it possible to begin developing a fuller understanding of the impact of robot design on the overall mission. In a world of semiautonomous robots, human decision-making is still a critical part of achieving mission goals. In current EOD team models—typically consisting of three to eight members, depending on service branch and mission type—potentially adding collaborative humanoid semiautonomous robots will be a valuable tool and one that can create a new social role within the group.

In addition to the role of the robot, the roles users have working with robots and the impact these relationships have on user perceptions of working with robots in teams is important when discussing the dynamic of human and robot in EOD work. Scholtz (2002 and 2003) defined several roles users may have working with semiautonomous robots in team situations, including supervisor, operator, mechanic, peer, and bystander.

In today's EOD work with current robot models, the EOD user's role is likely to encompass aspects of several or all of these interactions as Scholtz (2002 and 2003) originally defined them, including monitoring the overall use situation, operating, commanding, or maintaining the robot. As Scholtz stated, the overlap of some of these interactions may at times be blurred (2003). These roles, as Scholtz noted, have different interaction dimensions that affect the user in terms of goals, intentions, actions, perception, and evaluation when working with robots.

Robots

Explosive Ordnance Disposal personnel have a range of gear and tools available to assist their Render Safe Procedure methods. Here, the focus is on the robots as the significant device in the interactions being considered. The term *field robot* also has different definitions, and here the term refers to a robotic mobile platform that is semiautonomous and teleoperated, often used in a dynamic operating environment (Jones and Hinds, 2002). The primary user extends their own abilities to sense and maneuver by operating the robots at a physical distance. Thus, field robots principally interact with the operator from a distance, making the humans, in some ways, spectators.

Murphy (2004) specifically referred to the robots used in the contexts of space exploration, humanitarian rescue efforts, and military efforts as *field applications*. Using this reference point, field application domains then have two pertinent characteristics: first, these robots are subject to unstable environmental effects that can hinder the robots' stability and communication resources. For example, in a field situation, a robot may flip over in a ravine or lose radio communication with the operator in rocky terrain. The second characteristic Murphy assigned to field application robots is that their use is intended to keep a human from direct harm by operating at a physical distance from the primary user(s).

Robots used as tools on EOD missions share the same physical space as the human team members during transport and then are often the first line of contact sent out to investigate IEDs or relocate UXO to a distance safe for personnel. In the current model of EOD work, members collaborate at a mission's location, but may be somewhat at a distance from one another in the course of actions. For example, if a Team Leader is required to put on a bomb suit and use RSP on an IED because the work is beyond the scope of the robot, the Team Leader is then in communication with other team members from a geographical distance and technology provides the mediated communication for teamwork.

Situational awareness is a significant component of teleoperation tasks. A common problem with field robots is user reliance on video feeds and a lack of direct interaction with the environment, which hinder the operator's accurate understanding of robot location and stability (Casper and Murphy, 2003; Darken, Kempster, and Peterson, 2001; Lewis, Wang, and Hughes, 2007; Woods, Tittle, Feil, and Roesler, 2004; Scholtz, Young, Drury, and Yanco, 2004). The robot's operator must manage multiple cognitive tasks and dynamic incoming information to manipulate the robot's actions. Current EOD robot models use tracked or wheeled systems to maneuver in physical space. Finkelstein and Albus (2003/2004) reported that wheeled machines might operate on about 30 percent of the earth's land surface, while tracked vehicles can travel on about 50 percent. EOD robots are also required by their function and operating environment to move in spaces such as buildings and other human-made spaces. The advantage of using agile and stable-legged or even biped robots becomes obvious, then, with the ability to effectively work in a larger arena of surfaces and situations.

While robot models evolve and new robotics companies emerge, the principles of design are, at their core, similar. The ultimate goals for robot development in this arena are to ultimately build an effective, efficient, and flexible robust tool EOD personnel can use to safely disable, detonate, or remove unexploded ordnance with minimal human risk for people—to save lives. Currently, EOD robots are generally wheeled or tracked, and do not overtly resemble humans in appearance. Furthermore, they lack a humanlike ability to move truly agilely in all outdoor or indoor environments. A humanoid robot with biped legs and dexterous arms and hands could accomplish EOD tasks smoothly in difficult terrain via human operator control: they are able to move about in buildings, climb a rocky environment, operate existing machinery, and adroitly handle IEDs made by human hands. In fact, a 2004 Defense Advanced Research Projects Agency (DARPA) poll of U.S. military officers revealed their expectations that humanoid robotic infantry will be integrated by 2025 (Finkelstein and Albus, 2003/2004; Singer, 2009).

Supporting this poll, a 2004 DARPA-funded study of optimal robot forms reported "humanoid robots should be fielded—the sooner, the better" (Finkelstein and Albus, 2003/2004, p. 4). Because the DARPA arm of the U.S. Department of Defense (DoD) publicly indicates that they are exploring the options of humanoid robot design for use in EOD situations, these statistics and this study focus on American personnel in order to discuss ideas about how humanlike robots in U.S. military EOD teams may affect different aspects of everyday use.

Presently, robots commonly used for EOD work are either completely remote controlled or semiautonomous, and not easily categorized as humanlike. Figure 6.1 shows a TALON model typically used for everyday EOD work.

Figure 6.1   TALON IV®
Source: QinetiQ North America, 2012.

This TALON model is designed to be used for unexploded ordnance (UXO) procedures as well as rendering safe chemical/biological/radiological/nuclear/ explosive weapons (CBRNE), security, heavy lifting, and defense or rescue missions (QinetiQ North America, 2013).

Another robot used by EOD personnel in everyday work environments is iRobot's PackBot.

Figure 6.2   510 PackBot

Source: iRobot, 2015.

As seen in Figure 6.2, PackBots in some ways resemble the TALON, with a low body and tracked maneuverability. This robot can climb stairs, navigate narrow passages well, and was commonly referred to as an everyday EOD robot model used by participants in the study discussed here.

The heavy arms and claws on these robots are suited to tasks for inspection of UXO and their safe disarmament. In the past, some models of robots like these have also been weaponized for some tactical objectives. However, when used for everyday EOD purposes, these robots are meant to keep people safe by using them at a safe distance to interrogate devices. While one cost-saving goal from the military's acquisition standpoint might be to encourage the development of more multipurpose robots that can be used successfully across operations, the specialized nature of UXO interrogation requires specific abilities from its tools as well as its human operators. Some models may be used by troops not specifically trained in EOD work, but even the best-designed robot is not a replacement for the human expertise embodied within the EOD team. In addition, as with any technology, there is a demand for increasingly faster turnaround prototype-to-production times.

Ultimately, these robots are intended to be destroyed if necessary, a calculated risk when their role is to act in lieu of humans at close range. Within even a military context, this makes EOD robots unique. While all military robots may be ultimately dispensable, EOD robots are built and used with the expectation that the outcome of any task they undertake could ruin them beyond repair. This situation is not only preferable to any alternative that would put a human in harm's way, but is at some level not only an understood risk but a very real possibility. These are robots meant to carry out tasks in tandem with their human operators, but also to "take one for the team" as the target. These robots act as doppelgangers for the operator not only to carry out tasks and keep people safe, but they are designed and built with the knowledge that this thing used in such critical ways may not only wear out with use over time, but be blown up in front of the very people who use it every day. In the robots' current use modes and iterations, this risk may not be problematic for the users, other than forcing them into timely workarounds or the use of a backup robot when available. Yet over time, as robot design changes and the incoming people trained to use them have different sets of experiences—like growing up in a world full of everyday robots—additional human factors may be at play.

Operating Environment

Because the nature of military EOD work encompasses foreign and domestic locations and a variety of duties such as VIP protection and unexploded ordnance detection and defusion, it is impossible to say what a "typical" physical work setting or location is in this profession. Any environment can become an EOD operating environment. For those reasons, in this work, the term *operating environment* refers to an overview of the situations people are in, rather than focusing on one geographical site or specific incident type.

As an example of the complexity of describing any operating environment, there are five tenets to Army operations: initiative, agility, depth, synchronization, and versatility. Furthermore, there are specific descriptions of how these tenets relate to EOD, with more granular descriptions of the operating environment expectations as a whole (FM-105; 3-0, 2001). As defined in the field manual, any operational environment has six dimensions, with each factor affecting how the Army orders, collects forces, and carries out military operations. These dimensions are identified as threat, political, unified action, land combat operations, information, and technology. Written down and carried out, these are the formal Army descriptions of an operating environment and the expected operations within it.

It is worth explaining the overarching tenets in order to have a better idea of some of the expectations explained about the operating environment itself, as stated in the documentation. Even though there can be wide gaps between the formal definitions here and what actually occurs in the field, it is useful to understand how personnel are presented with an environment model.

Initiative is defined as the changeable state of actions that determine battles. Furthermore, the Army describes initiative as an "offensive spirit," used when conducting operations (FM-105; 3-0). Considered a holistic approach, *initiative* is the ongoing tenuous balance of effort used to compel the enemy to follow the Army's purpose and pace. For individual troops and leaders, there is an expectation to use individual actions within the framework set by command goals. Moreover, EOD are expected to anticipate needs and act prior to needs identified by higher levels.

Friendly forces that take actions prior to enemy maneuvers are the crux of *agility* as a concept. Specifically, EOD are counted on to be task organized and able to provide fast and concentrated responses to every situation. *Depth* refers to the use of time, space, and Army assets as extended to operations. EOD are counted on to act as support throughout a theater of operations, protect the commander's freedom of action, and extend flexibility and endurance of the larger organization and operational goals by reducing and eliminating ordnance hazards that threaten personnel, operations, facilities, material, and anything necessary to sustain combat operations.

Using all resources to maximize combat power is the process of *synchronization*. Activities such as logistics are coordinated throughout operations. EOD activities that control threats serve both the process and results of synchronization by providing "protection, mobility, firepower, security, and intelligence" (FM-105; 3-0). Moreover, there is an explicit expectation presented for EOD personnel to act in a coordinated way with all aspects of this process, and additionally includes the explicit requirement of anticipation, as well as expertise in the understanding of friendly-enemy interactions. The ability to move cognitively, emotionally, and physically from one mission to the next is the core of *versatility*. For EOD, this tenet implies they should be able to support a wide variety of operational, strategic, and tactical systems with ease and expertise.

From an opposing-force standpoint, one significant advantage of using weapons such as IEDs is that they are not mass produced but homemade, and are therefore unpredictable weapons not just because of their strategic placement, but because of their relative low-technology ease of construction, as well as their ever-present threat and overall effectiveness. While military training and formal descriptions of situations are crucial components to carrying out successful missions, the personal experience and knowledge of individuals is the advantage of each team.

Tasks

At every level of their job, in order to be successful EODs must be able to carry out individual tasks and *cooperative tasks*, or activities that take the coordinated effort of multiple people in order to complete the actions. Team structure and size are dictated by official military guidelines, but are also shaped by the nature of the work and, to some extent, the technology used, such as the robots. Team tasks differ from individual tasks in many ways, including coordination and communication between members (Nieva, Fleishman, and Reick, 1978; Naylor and Dickenson, 1969), and the dependence of team outcomes on performance by all members (Steiner, 1972).

Because of the nature of EOD work, every situation is unique, and therefore, even when following standard Render Safe and Disposal Procedures, the team must have a level of ongoing intrapersonal planning, decision-making, and negotiation, or *conceptual tasks*. Sundstrom, De Meuse, and Futrell (1990) describe military teams as "highly skilled specialist teams cooperating in brief performance events that require improvisation in unpredictable circumstances" (p. 21). This characterization appears to fit EOD teamwork in most circumstances.

The robots used by EOD teams as tools help complete joint tasks as part of the teams' everyday work and so, between training exercises, routine maintenance, and missions, for many human operators there is some level of human-robot interaction as a daily activity ranging in duration from a few minutes to many consecutive hours. However, not every team member works directly with a robot by operating it. Individual roles vary by leadership certification, as well as by specialized training and assignment. As mentioned previously, the standard number of people within each EOD team also varies for each military branch.

Using the Army's description of EOD Military Occupational Specialty (MOS) 89D EOD Specialist, specific duties are outlined at five "skill levels" (MOS 89D, n.d.). These duties are lists of *behavioral tasks*, things with physical behavior that range from the preparation of technical intelligence and incident reports to locating buried ordnance, radiological monitoring, technical mentoring of less experienced soldiers, developing new or modifying Render Safe and Disposal Procedures when necessary, and advising commanders when UXO are within their range of operation.

Team Leader duties encompass the first four levels of responsibility in addition to overseeing the team's safety and training, performing the diffusion of ordnance, and deciding how to diffuse explosives and what precautions to take. They also act as a liaison between the team and commanders and their staff. Team Leaders provide direct expert advice to senior Secret Service staff during presidential and other VIP details, and act as the team liaison with FBI, ATF, and civilian law enforcement when called to assist those agencies.

Conceptual tasks associated with EOD work, or socially interactive tasks like planning an appropriate course of action, negotiating approaches, and intragroup decision-making, however, are myriad, situated, temporally dynamic, and highly context dependent (Stewart and Barrick, 2000). One way to examine the conceptual tasks of an EOD team is via *interdependence*, or the extent to which team members cooperate and work interactively toward task completion (Campion, Medsker, and Higgs, 1993). Successful EOD teams require a high level of interdependence, where individuals rely on one another for information, materials, and reciprocal participation.

It is a fair hypothesis that introducing new technology into this dynamic in the form of robots with increasingly anthropomorphic design and with an ability to perform more complex tasks, communicate at a higher level, and exchange richer information with humans, will impact the current team structure and possibly a spark a reorganization of some behavioral tasks, if not conceptual ones. This new set of variables warrants continued research into how levels of human-robot interdependence may increase or decrease and change the current EOD team dynamic and, therefore, how it will affect implementing new training paradigms and achieving mission outcomes.

Chapter 7
Action Formation

In its relatively short history, the basis of much of current human-robot interaction research has been rooted in human-computer interaction and human psychology, often combining formal data gathering methods such as study participant self reporting or physiological and behavioral measures. Emotionally and culturally loaded things such as human experiences and social interactions can be analyzed formally, providing rich detail about our accounts of life. One approach to this type of formal investigation or study involves identifying informants who directly experienced the phenomenon in question, interviewing them in-depth, and examining patterns in their responses—the data—to reveal the internal structure of the interaction as an individual presents it. Social interactions can be described, which means they have an informational content in addition to whatever qualities or emotions are attached.

There are many models of the research process, most of them devised according to a series of stages. At this period in human-robot interaction research on user expectations, the researcher's goal is not to necessarily suggest a generalizable set of design heuristics for how to effectively structure a human-robot interaction, or to theorize about what all users in a defined group expect from dealings with robots. Rather, this study looked closely at a specific group of users in particular contexts. Examining certain groups of end users and their expectations about robots can provide our first insights into how other people in similar situations interact as a team or in collaborative situations with robots. Foundational work not only offers a description of what was explored, but is a basis for generating theory. Therefore, this study used qualitative methods of data gathering, with an emphasis placed on gathering descriptions of points of view.

This work represents one of the first steps in an attempt to disentangle the complex human factors side that are individual- and situation-based in EOD work, thus setting the stage to begin examining user-end effects and variables throughout the course of effective robot development.

Long-term research in this area can be used to help improve the troops' robotics training; enhance robot development specifications to mitigate mission-dependent risks; and improve warfighter and civilian safety in conflict environments, both foreign and domestic. More broadly, findings can be applied to the development of robots that are effective in a variety of human collaborative/team or training situations, especially in stressful conditions (for example, space, defense, and humanitarian relief).

This study conducted exploratory research on EOD personnel human-robot interaction practices rooted in individual human perception and experience.

Specifically, these human-robot interaction experiences were examined with two goals in mind:

1. Describe everyday human-robot interactions of this segment of users in terms of experiences, expectations, emotions, and actions.
2. Develop a holistic understanding of these users' everyday human-robot interactions.

The issues examined here have only recently emerged as an area for academic scrutiny; therefore, the nature of this work was exploratory.

Strategy of Inquiry

To address the goals of this research, there was a methods strategy appropriate for this study in order to comprehensively examine the complex set of user groups, their activities, processes, culture, and their interrelationships. Because the overarching strategy chosen to approach this work needed to support the understanding and subsequent description of EOD personnel experiences with robots in a rich way, it required the gathering of a sufficient amount of data to address the guiding research questions posed:

1. What are the activities, processes, and contexts that influence or constrain everyday EOD human-robot interactions?
2. What human factors are shaping the (robotic) technology?

These questions are compatible for research methods that focus on understanding and describing a phenomenon. Lincoln and Guba (1985) identified some tenets that scaffold qualitative research inquiry, and described these beliefs as follows: "Qualitative research assumes that there are multiple realities—that the world is not an objective thing out there but a function of personal interaction and perception. It is a highly subjective phenomenon in need of interpreting rather than measuring" (p. 17). This way of looking at people examines whole, complex systems and does not attempt to reduce findings to linear, causal relationships.

Qualitative research methods are often used when the results are intended to provide detailed description about context, activities, participants, events, and processes. They are about describing the phenomenon, rather than focusing on outcomes. According to Domegan and Fleming (2007, p. 24), "Qualitative research aims to explore and discover issues about the problem on hand because very little is known about the problem." Patton (1990) stressed that qualitative methods are "particularly oriented toward exploration, discovery, and inductive logic" (p. 44). Therefore, since very little is known about this specific area of research, this type of research method strategy is designed to be inductive and discover how people make sense of things and interpret the world around them.

These methods of data collection and analysis supported a higher likelihood of reaching an in-depth understanding of the complex factors that make up EOD human-robot interactions. In short, this study was not an attempt to quantify findings or find statistical significance but to use nonmathematical procedures to shed light on complex, internally located phenomena.

Individual Voices

Exploratory research does not always attempt to find a group of representative people across a population, but rather to follow an inquiry with people who have acknowledged experience and insight into a field. Therefore, this study used a type of nonprobability sampling called *purposive* (or, alternately, *purposeful*) sampling.

Participants in this study had to meet several requirements in order to be involved. Criteria for participant inclusion included:

1. Prior or current service in a branch of the U.S. military.
2. Training to work with robots in the field.
3. Working experience with robots in a military field setting over a period of time.

In addition, potential participants were deemed ineligible if their experience with robots was limited to drones, or fully autonomous robots, or if they only had experience with robots outside a military setting.

In total, the questionnaire and interview sources of data were obtained from a group of 23 EOD personnel. As an investigative study of EOD personnel-robot interaction in this domain focusing on personal experiences, the goal was to target a sample of EOD personnel that were self-identified long-term semiautonomous robot users, and had specific training to work with robots. Therefore, identification of this group of participants was key, and so the recruitment approaches targeted individuals with certain characteristics.

Closely tied to the question of which people to sample is the question of the size of the sample. In terms of data collection, *theoretical saturation* (Glaser and Strauss, 1967) is the point when no additional data is found in the process and analysis can begin. As a result, in research similar to this, it is rare that a specific number of participants are prescribed in order to produce sufficient data. In this study, 23 participants were determined to be a reasonable sample size because during data collection and analysis, no new information was found using the research methodology described in this chapter.

Narratives as Data

Data for this study were generated from two sources: questionnaires and semi-structured interviews. The questionnaire used here contained both closed-ended questions with limited response options, and open-ended questions that encouraged the participants to write in their own response. The advantages of open-ended questions include the possibility of gathering unanticipated data from respondents and gathering information from people in their own words (Fowler, 1988), which is in line with the guiding principles of discovery in this study.

The data from the questionnaire, combined with the information gained from the interviews, provided enough rich material to describe and analyze knowledge about the participants' interactions with and expectations of robots prior to EOD, and then from EOD training through their careers.

For researchers to understand the experiences of the study participants, they must establish and cultivate trust so that the participant feels safe enough to share their story (Charmaz, 1991). In addition to gathering data, the questionnaire was the first substantive interaction between researcher and participant, going beyond the initial administrative processes of recruitment and introducing the research, and providing the initial opportunity to establish rapport. Rapport building is critical in order to build a research relationship that will allow the investigator to access that person's story and to facilitate participant disclosure (Goodwin et al., 2003). Since the stories of the participants in this study have particularly sensitive contexts associated with military settings, some of the stories disclosed were self-identified by the participants as not having been shared before to friends or family due to personal trauma associated with some aspects of the experience. In an effort to withhold stories to protect friends and family from the danger associated with EOD work, some participants explained they had not gone into such detail before when sharing some of their experiences. On occasion, participants also identified concerns disclosing any secretive nature of a mission, and so did not volunteer details or withheld identifying details they felt were worthy of omission. Consequently, probing for information related to these experiences was not always an easy process. In addition to its usefulness for gathering information, the questionnaire had an added benefit of introducing the participant to the research process in a way that could potentially be less emotionally triggering than the interview that followed, framing a non-threatening space for the study and setting the tone for the interaction between researcher and participant.

Interviews offer a rich way to collect information such as participants' attitudes and experiences after observing human-robot interaction. Commonly, interviews are used to discover things that cannot directly be observed. The dynamic and interactive aspects of a semi-structured interview allow for follow-up and clarification on interesting, relevant, or significant points. In these types of interviews, there are topics identified by the researcher to explore, but not all questions are designed and phrased ahead of time. It is also a less formal approach, offering flexibility in a familiar conversational style between two people.

The interview structure applied in this research permitted spontaneous comments; however, where the participant did not spontaneously describe issues of interest in-depth, conversational probes were used in an attempt to engage the participant to additional narrative about the issue. This structure allowed the interview participants to speak in a detailed way about their experiences, ask questions, rephrase for understanding, and digress to related topics.

This research is exploratory in nature and focuses on discovery. In that vein, this study did not intend to test the generalizability or predictive power of a preliminary conceptual model. Instead, it collected data through a variety of techniques and then used inductive analysis to identify and characterize patterns of behavior, dimensions, and interrelationships in the phenomenon.

Human Dynamics of EOD Collaboration

Throughout the analytical process, the goal was to identify data that responded to the two research questions:

1. What are the activities, processes, and contexts that influence or constrain everyday EOD human-robot interactions?
2. What human factors are shaping the (robotic) technology?

In order to provide a richer description of each participant's background, during data analysis questionnaire findings were used in tandem with each corresponding interview transcript to acquire additional insights about individual experiences. It is also useful to look at the questionnaire findings apart from the interviews to understand basic characteristics of the group. Therefore, prior to presenting the results of the data analysis that resulted in the detection of categories and themes, it is helpful to identify baseline characteristics of the 23 individuals interviewed.

All 23 participants completed the questionnaire prior to the semi-structured interview. Of the 23 participants, 22 were male and one female. Their ages ranged from 22 to 49 with an average 34 years of age. Participant membership in the five branches of the United States Armed Forces were as follows: 15 Army, two Navy, one Marine Corps, five Air Force, and zero in Coast Guard. Five participants also identified themselves as serving in the National Guard and one in the Air Force Reserves and U.S. reserve military forces, in addition to a primary branch.

The number of years served in the military ranged from three to 28 years, with an average of 13 years. Two participants volunteered the information that they were retired from service; one participant reported he was scheduled to retire within the year. The U.S. Department of Defense does not make demographic statistics for EOD personnel readily obtainable to the public, so it is not known how typical this sample is compared to the entire population within the military.

In order to get a fuller picture of participants' knowledge and expectations of robots in general before their standardized EOD training and individual military experiences, the survey asked about their interactions, if any, with robots prior to military service. To get a fuller understanding of participants' exposure to robots prior to EOD work, the response options encompassed cultural media exposure via science fiction and toys as well as interactions with real robots. Ten participants claimed prior robot interactions, spanning the response options, but with most indicating exposure to robots prior to their military experience through toys and/ or science fiction.

A second question related to prior experience with robots was formulated to clarify any context in which participants had encountered or interacted with robots, with an emphasis on the workplace. The ten participants who had claimed encounters with robots prior to military work—including non-work environment interactions—elaborated on the context of their exposure to robots in this response.

Examples of the more detailed responses included:

1. Remote-control toy; e.g., programmable remote control (R/C) tank.
2. Engineer experience designing/building automated assembly line equipment.
3. Undergraduate college course on industrial robotics.
4. Tactical robots.
5. Bomb disposal robots (Talon, PackBot, Andros, Vanguard) with work experience.
6. Home kit robots.
7. Roomba (a home-use vacuuming robot).
8. Sci-Fi books, movies.

Yet another attempt to gain insight into participants' opinions about robots in general was to ask them to define the word "robot." This question was asked in both the survey and at the end of the interview. The goal of repeating the question was to give the participants time to reflect further on their idea of what a robot is and, if applicable, refine or expand their first response toward the end of the interview process.

In the response to the survey question, without exception, participants used one or a combination of the words in their definition of a robot: *tool, device, system, machine,* or *(electro)mechanical.* The interview responses to the same question bore similar patterns, with occasional extended elaboration (see Appendix A for verbatim responses from the questionnaires and interviews). It is interesting culturally to note that there is no one definition of a robot provided by the study participants, even among people (like EOD) who are heavily trained and indoctrinated with similar training as a group and are members of the same group. In many cases, the same people could tell you the verbatim definitions of other frequently used military jargon almost by rote. Yet their definitions of *robot* come straight from a personal space of experience, emotional and not necessarily only embedded in training.

In addition, right now, these people all used a variation of the words "tool," "machine," "mechanical," and similar non-organic words to describe a robot. This finding is important to note from a historical place in language and human-robot interactions, because based on what is known about the ways people have changed in our interactions with other technologies, it is likely definitions of a robot in similar scenarios will change over time, too. Ten or twenty years from now, people who work with robots in ways analogous to EOD might use the words "teammate," "partner," or "soldier" in their definitions of a robot because the technology changes, the ways people use the technology change, and the ways the technology becomes embedded in culture have changed.

The two conceptual themes that emerged from the data are the *Human-Human Interaction Model* (HHIM) and the *Robot Accommodation Dilemma* (RAD). From the interview data, analysis revealed distinct patterns of beliefs, values, and strategies connected to successful EOD human-human interactions. This set of human factors in EOD culture is referred to here as the *Human-Human Interaction Model* (HHIM). These same factors identified by participants as part of successful human-human collaboration were rarely reported as existent in their everyday human-robot collaborations. Rather, the model for human-robot interaction that emerged encompassed the conflicting emotions and expectations of interviewees about robots in a set of consistent themes that are collectively referred to in this work as parts of the theoretical concept called *Robot Accommodation Dilemma* (RAD).

The first theme, HHIM, is the social framework the participants engage in to communicate and connect with other members within the EOD group at an organizational and team level. The participants described their initial struggles with the period of intense specialized training at The Schoolhouse, and how they successfully navigated both the intellectual and assimilation challenges.

During The Schoolhouse training, they connected as new members to the EOD group by virtue of the common training and centralized geographic location with other EOD trainees from all military branches. They then participated in ongoing learning activities and everyday actions of their job as part of a team that interacts with critical care toward achieving mission results. In order to adjust, adapt, and overcome personal problems, challenges, and distractions, participants needed to make a concerted effort to internally mediate all those frustrations and fears in order to progress from being uncomfortable in a new environment of EOD work to comfortable learning and working together toward common group goals. In order to be successful in EOD work, participants identified crucial factors in human-human collaboration, which were further parsed into the three categories of *beliefs*, *values*, and *strategies* during the analysis process.

Although there were no specific questions about the best practices of human-human communication in EOD work, participants raised the subject without explicit prompting. The concepts regarding successful human-human interaction emerged from the interview data in patterns defined in this work as *beliefs*, *values*, and *strategies*. During analysis, *beliefs* were defined as a state in which an individual holds a proposition to be true, often with examples alluding to ways of thinking about their role as EOD. *Values* refers to a pattern of participant responses that reflect participants' sense of right and wrong or what *should* be, according to their experiences and the implicit and explicit demands of EOD work; values tend to influence attitudes, behavior, and how individuals act. The term *strategies* refers to patterns of responses in the data that describe concrete actions participants use to do their work successfully. Some of these actions may be formally learned in training, while others are informally acquired through cultural negotiation on the job.

Throughout participant responses, these beliefs, values, and strategies emerged naturally into the idea of EOD self-identity as more than a sum of competencies, but beliefs about their individual and collective ability to exercise their competencies, especially under challenging circumstances. An important overarching concept about beliefs, values, and strategies is participants' confidence that they can adapt their expert skills and core competencies effectively in the dynamic and challenging conditions of their job.

As seen in the interview descriptions of EOD training, personnel are given the situations to develop core competencies that come from modeled behavior, job experience, analytical reflection of situations, and performance practice under high physiological and emotional arousal. Although EOD missions have many different variables, the participants consistently expressed an overall strong sense that their learned skills, experience, and resources used to achieve RSP was portable between work circumstances. The voices of the interviewees are the ultimate data. The quotes included here are representative and help to richly illustrate what and how the interviewees think and feel.

All the participants here are presented using pseudonyms in order to protect their identity. It was also ultimately considered prudent to change the names of the robots referred to in individual narratives. Since the EOD community is relatively small in relation to other groups within the military, the robot names were determined to be too revealing a detail about participants to leave unchanged in published results. Additional potentially identifying information, such as references to specific geographic locations, has been redacted from the quotes. All other aspects of the transcription excerpts are intact.

Believing the Work is Unique and Challenging

The first questions of the interview focused on each participant's choices surrounding their entry into Explosive Ordnance Disposal as a career choice. The variety of answers demonstrated a cross-section of individual reasons for choosing EOD work, ranging from monetary motivations for sign-up bonuses to responding to recruiter suggestions. However, two overarching patterns for underlying motivation for joining and enjoying EOD as a career were revealed: (1) ascribing uniqueness to EOD work, especially compared to other groups in the military, and (2) the implicit and explicit challenges of EOD work. The ideas of *uniqueness* and *challenge* are worth discussing separately in order to differentiate between the closely tied concepts; these ideas are defined differently in this research in order to distinguish each as the participants implicitly do in the context of their responses.

Although a challenge might be considered something negative to a person reading the responses out of context, participants most often described the idea of a challenge in positive terms. Participants' choice to use the word *challenge* may come from common military cultural language, but in participant responses, this word consistently described something associated with achieving positive outcomes from difficult situations. It was highlighted as especially important to their individual identity and to their identity as a member of the EOD group.

The word *unique* was taken from the participants' own interview language because it was used repeatedly to describe interviewees' perception of EOD personnel and work, including their initial attraction to the field and their ongoing satisfaction with the career. The idea of uniqueness has to do with participants' self-identified desire for possessing a distinctive career, either in general or within the military, and the belief that EOD work fulfills their definition of uniqueness. A common pattern emerged in the data of participants describing this desire for a unique career, closely followed up with a statement about their belief that EOD work and culture was a good career fit because it matched their existing unique or unusual personality, intelligence, or social characteristics.

Furthermore, this belief of EOD of *uniqueness* was frequently something that rested on a mythos surrounding the image of EOD personnel as special, interesting mavericks and individuals. Patterns emerged from the data that this mythos rested on two major points: (1) EOD work emphasizes *individual* (as well as group) evaluation of risks, and places an emphasis on effective intergroup communication, and (2) participant interactions with EOD personnel helped form this image and influenced their opinion of the job before they joined. This former concept of retaining or expressing *individuality* in a military organization was a significant feature of uniqueness when participants explained their initial attraction to EOD work, and after becoming a part of the culture. The feeling of EOD as a unique group is reflected in this quote:

> EOD used to be a lot like Special Forces, you know, this small group of guys that nobody knows what they do, and they sit in the corner of the base. You know, they pretty much do their own thing. Everybody leaves them alone. (Axel, 26, SGT, Army)

The idea of EOD as separate, but still associated with the military, nods to the dual nature of this type of uniqueness as a desirable thing. The idea of individuality within the military is also illustrated in this response:

> It's not the usual day-in/day-out rigors of the military. Whereas, an infantryman will do the same job, everyday, by the book ... practice the same movement, by the book, everyday ... one of the aspects of our job is to be unique and individual while staying within a certain set of guidelines. We're able to approach everything differently, use our imagination to defeat a problem. Working in small teams, it's definitely a distinct challenge. It's fun. (Irving, 31, SSG, Army)

Irving's quote is typical of participants literally using the words *unique* and *challenging* in the same coded segment, as well as figuratively tying the concepts of uniqueness and challenge together.

During his interview, Roy described the "air" of EOD work by alluding to the latter part of the EOD mythos when he referenced the Oscar-winning movie *The Hurt Locker* (2008). He explained that the film is a fiction-based cultural touchstone and a model for how non-EOD people may regard those in the field, whether actual EOD personnel consider that portrayal realistic or not.

> I chose EOD for the money, actually. When I was 16 I really didn't, I'd taken the entrance test to the Air Force and I qualified for all of the jobs. I really did—I certainly wanted to have a special—I wanted to do something unique ... But one of the things about the EOD career field was there was a high wash-out rate which intrigued me ... And I—my dad is a reservist in the Air Force as well, and when I had gone off for one of his reserve weekends, they had a family picnic up there, and I remember seeing those EOD guys, and seeing just their—I've never really had a really relaxed attitude compared to the rest of Air Force personnel that were there. They laughed a lot and they joked around and when you talked with them you could—I mean, they could put the humor aside and you could, you could tell that you were talking with someone, you were engaging someone who could engage back with you on a pretty high intellectual level. And so that was really rewarding for me.
>
> That—and in addition to going through our preliminary course and going through our school we started out with 30 from our initial preliminary course. And of that 30, only three of us graduated. And so the longer that I went into the career field, the more arrogant I became—you know, the taller I walked around. And there was an air about it.
>
> When people heard that, you know, this is what you did, amongst the Air Force—so before the movie *The Hurt Locker* came out a lot of people didn't understand what we did. They didn't—they'd never heard the term "EOD." ... They didn't really know what we did, and so as far as walking around on base we got away with a lot of stuff just because no one really knew what we were supposed to be doing. So that helped feed into that, "Hey, we're EOD and we're the smartest guys on base." (Roy, 27, SSG, Air Force)

Roy describes a physicality here not just in terms of fitness, but also in a way that defines members on sight to other EOD by a unique overall demeanor and being different from others. Hector told his story of discovering EOD work this way:

> I was in the lead vehicle, so I was the one findin' all the IEDs, and dealin' with 'em and then workin' with the EOD, and gettin' em there and recording. And I was really impressed with 'em, 'cause I didn't even know what EOD was before then. I ... had been in for eight years at that point and I didn't know. So I was really impressed with 'em. They're sharp, and everybody was squared

away. They're—in the military, you always have guys that I call turds. They're just—they're there 'cause they couldn't work anywhere else, couldn't have a job anywhere else. They didn't have any of this. That was what really impressed me, and it's a brotherhood, too. They take care of each other. So they were the best and the brightest I'd ever met. So I wanted to go do that. (Hector, 27, SGT, Army/National Guard)

Hector's reference to EOD personnel as "sharp" and "squared away" further emphasizes his impression of them as being observably and impressively different from other troops.

During the interview process, when probing more deeply into these ideas, participants also identified the unique nature of the job as one that was preventative in its goals and results rather than destructive. There are other MOS within the military with similar goals of prevention as opposed to destruction, yet EOD work combines all of the elements participants identified as desirable, such as the ongoing rigorous physical, mental, and emotional requirements and challenges, as well as a culture that requires individual input in order to continually improve the training and skills.

Another example of this aspect of career uniqueness is reflected in this interview exchange:

Brady: I guess the reason I went into being an EOD tech was because it's one of the few places in the military where we get to help people instead of shoot people ... It's my job to make the bombs go away, instead of dropping bombs, you know? So that was my motivation, pretty much. I'd talk to people who had been techs, and they told me it was a little more relaxed than the typical Army life ... a little more closer, family kind of feel to the organization than just a "Do this, do that, yes, Sergeant, no, Sergeant."

Researcher: That's interesting you said it seems a "little more relaxed." Can you tell me more about that?

Brady: Well, I think it's because it's such a high-stress environment ... when you're actually out there doing the job and interacting. When you're not on the clock, the other things seem tedious even though they may seem high stress to other people. Comparatively, it's no big deal to us. So if, I don't know, for instance, a normal infantry guy tells his Joe to go do something to do, he has to do it. You know, the Sergeant starts making him do pushups, the guy's gonna be stressed out about it. Whereas, it's just pushups. So what, who cares? You know? (Brady, 28, SGT, Army)

As Brady's quote also illustrates, it is important to point out that participants described the uniqueness not merely from an ego-centered concept of betterness, but also as the perception that they make a tangible and positive difference to others through their work. Simon's explanation is typical of many participants' overall positive feelings about EOD work:

> It's something that you—that you're doing that very few people can do, and you know at the end of the day when you walk out of the office and you shut the lights off, and you lock the door, you made a difference. (Simon, 49, MGySgt, USMC)

Simon aptly sums up the everyday motivations to be part of EOD that centered on more than individual aptitudes, but also the ability to apply skills in a way that keeps military personnel and civilians safe or safer in hostile environments and situations.

Valuing Close Relationships with Peers

Related to the idea of uniqueness, almost all of the participants referenced feeling a sense of *brotherhood, fraternity*, or *family* (their words) among EOD personnel, regardless of military branch. Earlier, Hector spoke of his impression of the interrelationships of EOD personnel as a strong part of his continued attraction to the work: "That was what really impressed me, and it's a brotherhood, too. They take care of each other." A pattern emerged from the participants' stories about the significant value they placed on this type of close bond with other EOD, as Simon stated this way:

> One the few things with the EOD, the EOD program does, is we talk to each other. We have—everywhere that you go, you have a place to stay. You have a family. If they're wearing an EOD patch, they're your family. Maybe kind of dysfunctional, but they're still family. (Simon, 49, MGySgt, USMC)

Simon explained that the bonds of group membership go beyond one-on-one relationships with the people he knows, and extend to anyone he identifies as part of the EOD "family." He also emphasized the practice of "talk" and communication as a scaffold for the close-knit feeling between members. This in-group sense of belonging surfaced as a tangible thing valued by the interviewees.

The other side of the inclusive nature of the concept of a family is, of course, the idea of exclusivity of those who are not in the same group. Many participants described their first experiences in The Schoolhouse training as the place where they established themselves as part of a cohesive group. Those students who do not have the aptitude are excluded by the demands of the process, while those who remain are included in the elite circle of qualified EOD graduates. Hector described the importance of the training and the subsequent process of narrowing down the group membership, and how this experience differed from his Army training experience prior to The Schoolhouse:

The thing about as an engineer, it's—it's a big Army course, and you know you're not gonna fail. Nobody fails it. Even guys that are pretty much not smart at all don't fail it. They get 'em through, they'll hold their hand, whatever you gotta do. And so I hated that. You know, if you pass the test it should be because you passed the test and did good. But EOD, it's not like that. If you fail, you're out. You're kicked out of the school. You drop, whatever. I really like that. You keep all those guys that just can't hang—even if they're good guys, they could get one of your buddies killed. And so I didn't—I don't wanna work with people like that. (Hector, 27, SGT, Army/National Guard)

As seen in the above example and throughout this work, EOD Schoolhouse training and everyday EOD experiences are well documented as rigorous and demanding, mentally and emotionally. Based on this knowledge, probe questions were developed to identify issues experienced by the participants associated with the rigors of The Schoolhouse training and day-to-day EOD work, and interviewees were asked to describe how these conditions affected them from an emotional standpoint.

Marcus summed up his feelings about the initial formal education at The Schoolhouse this way:

It is academically the most challenging thing that I've ever been a part of ... . But, it was exceptionally rewarding and it has kinda given me a perspective and a outlook on things that few people in the military share, and I feel kind of a—for lack of a better word, "elite" and unique and yeah, we're a pretty big deal. Yeah. It was—it's fantastic to be a part of this fraternity. (Marcus, 32, TSgt, Air Force)

A pattern arose of participants describing this initial winnowing process via The Schoolhouse training experience as affecting the process of social cohesion (Kirke, 2009; MacCoun and Hix, 2010) among new EOD. The U.S. military is composed of many formal groups that nest in a strict hierarchical chain of command, such as this Army example: Army>Corps>Division>Brigade>Battalion>Company> Platoon>Squad.

Other formal groups, such as EOD, are subsets that are part of these larger hierarchies. Participants repeatedly identified the The Schoolhouse as an important part of their feeling that they were part of EOD culture at a group level. In The Schoolhouse, the social structure of EOD is built in part by common training, as new group members learn the formal technical processes and rules for behavior, such as Render Safe Procedures, which are uniform across branches.

In addition, The Schoolhouse is where EOD begin to learn some of the overarching conventions of behavior, such as ongoing verbal communication during missions. The attitude of uniqueness appears to emerge for EOD personnel during this first formal training period, and with that, the first feelings of individual troops evolve into a "we," or the sense of belonging to EOD at this level prior to being assigned to a smaller unit. This subset of personal identity is part of the social structure interwoven with the operational structure of the military.

The sense of family or in-group association as they became immersed in their new role within the military relied in part on establishing bonds with peers who were successfully navigating The Schoolhouse environment. The shared common learning and survival experiences affected participants in a way that was portrayed by participants as challenging and ultimately an overwhelmingly positive experience. Patterns of specific strategies for working successfully in the EOD field emerged; in a comment typical of how many also expressed their lasting impressions of The Schoolhouse, Jed shared his memories:

> It was ... it's classroom for about eight hours a day. Well, not classroom. It's training for about eight hours a day. And then generally between one to three hours of study hall at nighttime. So it's a lot of information. It comes at you very fast, and there's a lot of tests and performance reviews. You go through that to make sure you're picking up all the stuff. So, yeah ... what I remember about it is just ... and it almost was actually a really, really good time 'cause you're with a good group of guys, day in, day out, every day. So you get a pretty tight bond with your classmates, as well as just trying to swallow all the information they're pumping at you as fast as you can. (Jed, 41, SCPO, Navy)

Strategies Learned Within the Culture for Survival

From the data, specific strategies emerged that interviewees used for managing stress and cognitive overload, such as the compartmentalization of emotion and the purposeful knowledge exchange between team members, in order to make life-altering mission decisions. According to participants, these strategies were learned through a combination of formal training and their observation of EOD culture and practice.

Regardless of what year the participants had gone through The Schoolhouse training, participants described the experience as *stressful*. Perhaps more significant is that the stress was associated with positive feelings. As Jed noted, the training was "stressful," but "you keep doing what you have to do and you deal with the stress afterwards." The word *stress* was frequently attached to positive contexts or outcomes, especially in terms of something that can be channeled toward a goal, or otherwise purposefully ignored in order to focus on the tasks.

Another participant, Quinn, explained how he believed stress successfully triggered his training reflexively in a dangerous situation: "At that time I view that as a good stress, because training takes over" (36, TSgt, Air Force).

Participants also identified close ties between job satisfaction and the need to be mentally and/or physically challenged by the job. Roy explained his desire for challenging work, and how without that level of career engagement, he felt unfulfilled for a period of time:

The first four years of being EOD … this is before we were heavily engaged in [location redacted]. My supervisor that I had was really awesome and I was really proud of what I was doing. They kept me constantly engaged, constantly challenged. When I went to my second relief station I went to a much smaller shop with a much different mission … .we went on one response in two years. One documented response in two years, Stateside, and a very small range appearance which was basically glorified trashman. And so I really struggled with a sense of job satisfaction because basically, I would come to work in the morning, and for the last ten months of my active duty career I didn't go on one single fall Stateside, and so I really struggled with that satisfaction. (Roy, 27, SSG, Air Force)

The connection between *challenge* and *stress* was often explained by participants as something to be (purposefully or innately) balanced by identifying and pigeonholing these states of being as separate conditions in order to survive the unusual circumstances associated with EOD work. Marcus explained his experience observing emotional compartmentalizing among his peers, and his insight about the emotional separation he uses to manage stressful facets of his work:

We have a unique ability to overly compartmentalize. So the—the job is certainly stressful, but we tend to—I have noticed that I tend to feel the stress after something wiped out, rather than while it's going on. I'm kinda detached from the fact that, you know, it's, such as a life or death situation and rather just kinda focus on what we've got goin' on at hand. However, afterwards could be, you know, a little unnerving of thoughts. (Marcus, 32, TSGT, Air Force)

Compartmentalization of emotion was one of several strategies the interviewees identified as part of their negotiation of everyday work. Examples of compartmentalization in order to work through challenges were discussed particularly often in relation to boots-on-the-ground work after The Schoolhouse graduation. Participants sometimes explained that they felt it probable that some of this emotional compartmentalization was learned behavior from military training developed and applied with that outcome in mind, in order for people to effectively work though missions. However, these participants were also very aware of their ability to separate emotions (e.g., fear) from work while in the moment, and reflected on this tendency, sometimes attributing it to their own personality in addition to something learned via training or experience.

In this quote, Simon reflects on his ability to separate at-hand tasks and decision-making from dwelling on the possible long-term negative outcome of his actions:

It was my job. It's really weird. You don't think about the outcomes. Your concern is what's in front of you. I never thought about any of that stuff until afterwards. 'Cause it doesn't really do you any good. 'Cause just thinking about

> it, you can make a mistake and usually, in my line of work, you don't get to make a mistake. (Simon, 49, MGySgt, USMC)

Rashad explains the need to separate immediate personal safety concerns from the more global mission-dependent tasks, and how and why he purposefully separates emotion from work if he is able to, in order to survive:

> I was not very affected until the death of a friend, but ... for the first 12 months of our deployment, I was a little bit concerned in my lack of emotion about ... They—our job is gory. We get into ... I get around dead people and things like that. Aside from a U.S. casualty or something, which is emotionally difficult, a dead terrorist or something like that does not negatively affect me. The stress-wise, as far as ... worrying that an IED is gonna blow up or something else, really wasn't there so much. I just kinda took the practical view of: "Worrying about it is not gonna help anything. I just need to be on my game. I need to be able to relax."
>
> I think that there's kind of a ... I wouldn't call it a rush so much of what you think about, but the ability to go from inactive—not doing anything —to emergency situation quickly, was something that I had to do on a frequent basis. You would drive for hours and hours every day and nothing would happen, and then all of a sudden, boom! Everything's happening really quick and it's sort of a ... switch flicks in your mind and it's ... it's work time, and every single ounce of your brain is focused on doing what you need to do, and everything else just goes out the window almost. As long as you have something to do.
>
> Now, at other times I've been sitting in the backseat of a vehicle while we're being ambushed and I have nothing to do but see if they got our ammunition. And then your brain starts to wander because you don't have anything to do other than hope that an RPG doesn't hit your window. But when I'm doing my job, I'm very focused on my job and making sure that everything else goes on and I'm thinkin' a mile a minute about 2,000 different things. (Rashad, 26, SSGT, Army)

It is important to emphasize that this type of intentional and unintentional moving of emotions into a box—to be examined (or not) at a later time—is a different process than the act of being immersed in an intense intellectual focus on the tasks at hand, as Rashad clarifies in the above quote. In fact, participants generally discussed their need to focus to an almost hyperaware state on critical tasks involved for RSP of unexploded ordnance, rather than spending little attentional effort toward their job in an auto-immersion mode.

Closely tied to the ability to emotionally compartmentalize was the participants' recognition of their private thoughtful analytical process, an action often compared to an ongoing internal narrative, as in this example, also from Simon:

Well, you think of what can happen. Certainly, nobody wants to die. I wasn't afraid of getting hurt but you ... you wonder about, you see people that are missing an arm, you know, digits, fingers, hands, you know, arms, whatever. And you know, you wonder, you know, what life would be like that. Or, you know ... you know, if ... the worse case, you're never gonna know, 'cause you'll be dead. But just those little things. You know, the things that you're concerned with when you're going down there is keeping your eye on what—what's there. Running scenarios through your mind: What do I do if this happens? What do I do if that happens? You know, what do I do with, you know ... I come down here and this is something really opposite of what we thought it was? Can I, with what I have with me right now, can I make those changes in that scenario? And I've got to make that decision, and I've gotta make it now. (Simon, 49, MGySgt, USMC)

Throughout the interviews, participants explained their conviction about the importance of decision-making and communication skills in order to work in EOD. Specific strategies used to communicate effectively between EOD group members emerged from the interviews. Identified as one of the most critical of these communication strategies is what was referred to in this data analysis process as *purposeful knowledge exchange* (PKE). Elements of PKE include intergroup (1) problem identification, and (2) negotiation of choices. Participants acknowledged that there are specific required purposeful knowledge exchange activities in their routines, such as the verbal description of RSP to teammates prior to attempting the processes, or the writing of incident reports.

In this quote, Simon accounts for the ongoing learning process from intergroup negotiation of choices this way:

If I can get there by doin' this, and I can do it faster and I can do it more efficient, and I can make sure it's correct ... Then the other three guys will either agree or will agree to disagree. OK. You do it that way, but I don't think it's gonna work so ... Yeah, OK. Noted. Then when I go in, it works, OK. Yeah. Put that in the memory bank. That'll work. If it doesn't work, do not say, "I told you." OK. "Here's where I think we went wrong." And there are no real right answers. And we always used to tell people, "If you don't have a thick skin, you need to find—you need to go buy one." (Simon, 49, MGySgt, USMC)

Certainly, intergroup communication is part of EOD training. The willingness to listen and learn from peers and apply new knowledge was repeatedly expressed by interviewees as desirable as well as a necessary, ongoing, and collaborative action between individuals and the team.

And so, and generally, what everybody has is a major and a minor. And you know, you have, so you get used to the one part. You have ground, you know, like hand grenades and landmines, and then you have the stuff they shoot out the big guns. Then you have the air ordnance, and we have underwater ordnance. Then you have biological, chemical, and nuclear. And then you have improvised biological, chemical, and nuclear. And so you try to major and minor. And so I might be really good at one thing. Then we go somewhere and we're like, "Hey, Bob, what do you think?" You know? And I'll defer to that person, and just not really get in on it. So, it—it really just depends on what you feel you're—and so nobody's really good at everything. There's a couple guys who are just really good at everything. But, you know, you—you try to pick one and you—and it's what you feel comfortable with. Does that make sense? (Leon, 45, PO1, Navy)

However, purposeful knowledge exchange is not just a strategy used in the course of missions or as part of the formal processes of mission debriefing. Here, Jed articulates how discussing incidents with others in the group relieved stress, in addition to contributing to group knowledge through the analytical walkthrough of work-related tasks.

Researcher: When you say "blow off steam" … can you give me an example of that?

Jed: We would … yeah … we would do all of that. Kind of the … it wasn't a really regimented thing, it wasn't planned or anything like that but kind of a habit we fell into … was when a team came back we'd all get together, we'd talk about what they had, what they did. We kind of put it up for discussion, of like, "Is there anything else you could have done, is there anything you could see?" The guys that hadn't been out would ask questions and kind of say, "Could you have done this, could you have done that?" So we'd break the situation down, make sure that the team had done everything they could right, as far as they knew.

We'd try and come together on agreement on any improvements that could have been done, or things that could have been different, or maybe some risks that hadn't been seen at the time, and maybe in retrospect had seen, so we'd do that for just the training and improvement side of it and to keep everybody involved in what was happening 'cause as time goes on the tactics … the enemy tactics change so it's good to keep current on the things as you see it.

And in that … during that process, as you kind of go through everything and that's when the, you know, the jokes and the stuff and the laughing kind of come around and then … so you can … you start to make light of the situation once you have a good understanding of it; everybody who went through it is safe, this … and the improvements we have and they can start, you know, kind of poking fun of each other and laughing and having a good time with it after that. And then it just helps to kind of lighten the mood and keep everybody optimistic going through, you know, hard times. (Jed, 41, SCPO, Navy)

Jed's description of an informal mission debriefing as a way to manage stress via knowledge and humor illustrates how critical the verbal communication is between EOD, tactically and emotionally.

The second emergent conceptual theme is called the *Robot Accommodation Dilemma*, or RAD, and stems from participants' description of their experiences with EOD robots, which ranged from appreciation for the robot as a critical EOD tool, to frustration about robot technical abilities, to descriptions of the robot as an extension of the operator's self. The meaning of the words *accommodation dilemma* refers to two main patterns revealed in the data regarding participant human-robot interactions:

1. Regarding robots as critical tools, and the importance of thoroughly recognizing robot capabilities and limitations.
2. Defining robots as mechanical, yet still developing ways of interacting with robots as a technology (e.g., as an extension of self, humanlike, animal-like, or uncategorized "other").

Although there were a variety of reported experiences and opinions about robots used every day, these two categories of interview data formed consistent and significant enough patterns to be identified and explained here collectively as RAD.

When speaking of the robots they had used, participants shared an overwhelming sense of robot as something "mechanical" and a "tool." At the same time, there was a trend to explain the robot as an extension of self. In other words, there was little evidence of EOD genuinely mapping human-human emotions, affection, or expectations on to their robots as they would be expected to map them on to another human friend or colleague. Yet equally meaningful patterns in the interview data revealed that participants often described robots as an extension of self, or as a team mascot or zoomorphic entity, or referred to the robot using language or cultural conventions usually reserved for living entities, such as referring to the robot as *he* or *she*.

The concept of *self* is the core of the subjective world in which human beings function. Like other features of our individual social realities, the self is principally a process and not a static state of being. It has been theorized that body parts, ideas, personal possessions, people, and places can be all incorporated into a person's sense of self. Yet the concept of self-extension is not necessarily about an emotional attachment to something. Rather, self-extension implies that a thing has strong symbolic meanings associated with a person's self-identity and definition of self.

Participants also described using the robot as a teleoperated stand-in for the human user, then often leading to associations of EOD inserting themselves into the robot's existence as an avatar, a thing closely associated to the physicality of their bodies. To a lesser extent, some individuals described this sense of self as if they inserted the operators' personality into the robot, and claimed being able to recognize characteristics of other operators via their robot tactics and maneuvers. In addition, a significant pattern emerged from the data indicating that the interviewees viewed robots as useful tools, but with problematic technical limitations.

On one hand, because of robot usefulness in some situations, there was a consensus among participants that robots should be used instead of a human team member whenever possible in order to keep team members safe. In the words of Rashad (26, SSGT, Army), "The reason why we're using robots is because they're expendable." Yet the sense of self-extension into the robot combined with occasional frustration with its limitations as a tool combined into a seemingly conflicting set of emotions that compose the accommodation dilemma.

Understanding Robot Capabilities and Limitations

Although a clear understanding of any tool's characteristics is important in order to understand its capabilities as well as its limitations, across conversations with EOD the robots' limits were a source of constant concern. Over the course of her interview, Sarah explained a frustration for her as a robot operator:

> There's a lot of situations where you're dealing with it when you have a task to get done with the robot and you're trying to get it done quickly and it's not going as planned. The robots can be so … finicky, I guess. One second they communicate, the camera's working great, you know, you feel like you have it. Then you lose comms [communication] for two, three seconds and you're turned all around again. You don't know … you're disoriented down there. So there's lots of situations like that stick out. You just learn to take a breath and try to see what's going on. Stop for a second, turn the cameras, get oriented again where your robot is on and the position that it's in and start over again. (Sarah, 27, SPC, Army)

Anxiety about robot reliability was repeated throughout participant responses. Aaron (31, SST, Army) expressed his feeling about working with robots in this exchange:

> *Aaron*: But, I had a lot of issues with them not working at times … so, the whole thing about emotional relations with robots? The most common one I'd say, if we felt anything towards the robot, it would be anger and frustration.
>
> *Researcher*: OK. Tell me more about that.
>
> *Aaron*: Well, most of them, we just used a radio control system on them. And it would lose comms [communications] a lot. Occasionally they would just do random crazy things. There'd be times when you'd be driving downrange and all of a sudden, it just starts spinning in circles. Not really sure what's going on with that.

Researcher: You're saying you have no idea why?

Aaron: No. It just happens to robots occasionally.

Aaron's frustration was directly related to what he perceived as the unpredictable behavior of the robot, or its unreliability, and this was how many interviewees described their hesitancy about robots as things to be consistently relied upon.

Other concerns participants expressed over robot limitations were closely tied to lack of trust or lack of confidence based on their personal experience with robots.

> I didn't have much confidence in that robot because of the downtime. Every time we tried to use it, it would either be so difficult to use ... you'd just get it ready to go it would—either the batteries would die, the thing would make some maneuver that would make it swerve out of control and knock everything over. Bang into the wall or something like that. [laughs] I never had a lot of confidence in it in a real defensive situation. (Mino, 49, TSGT, Air Force)

Participants described many situations that spanned a range of geographic locations and a variety of mission conditions to illustrate their perception of the inconsistency of robot behaviors, and the associated unreliable performance of robots and their limited capabilities. In several cases, participants reported jury-rigging robots on the fly (e.g., using duct tape to secure a tool to the claw) in order to overcome specific technical limitations. However, interviewees overwhelmingly appreciated the robot as a useful EOD tool.

This tension between reliance on robots and recognizing their limitations was also explained frequently by participants during the interview process. Jeremy, an Army EOD Team Leader, shared how he finds robots a practical tool, but also described their technical limits in some environments.[1]

Researcher: What do you think about robots now?

Jeremy: I like them a lot.

Researcher: Why is that?

Jeremy: Well, because my experience in [location redacted] ... you know, being a Team Leader, if we didn't have a robot, that means I would have to go downrange wearing the Bomb Suit and risking my life. So a robot, it's great for

---

1 Jeremy participated in member checking as a way to verify the accuracy of the study findings. After reviewing this transcribed comment, he asked that his response be modified from the original statement referring to his experience with robots from "you always have problems with them," to less concrete wording, suggesting a change to "can" instead of "always."

being the eyes and ears remotely to look at stuff, manipulate items … you know, from a safe distance. So it's saved a lot of lives, for sure.

Researcher: Is there anything you don't like about robots?

Jeremy: Other than … you [can] always have problems with them.

Researcher: Tell me more about that. Can you give me an example?

Jeremy: Well, like sometimes they lose connection, you know, so you may have to go retrieve the robot. It may get stuck; um, again, you may have to go down there and retrieve it. Other issues we have are some of the tools we use, like shock tube to place a charge downrange; sometimes the shock tube gets tangled up in the robot and there's no way around it, you have to go down there and recover it. But they … I think they've advanced a lot. When the Iraq war first started, they were still using the [robot model redacted] … which, it's a good robot for certain purposes. I'd say more for stateside response. For insurgent vehicles at a fixed location. It's not mobile or transportable by any means; it's really slow. It doesn't really work for response missions in Iraq or Afghanistan, just 'cause it's so bulky, cumbersome, and slow. They've evolved into robots like the [robot model redacted], the [robot model redacted] … which, to me, are very functional and do the bare minimum for what we need to do for the majority of incidents in Iraq or Afghanistan. So, I'm glad we actually purchased those like in the '04 time frame. (Jeremy, 34, MSGT, Army/National Guard)

Similarly, other participants warned that overreliance on robots, like any technology, can cause its own set of problems, such as limiting a user's practice of alternative problem-solving methods when a robot is not available.

I feel it's—I look at it: it's artificial intelligence. And you only get what you put into it. So if the operator behind the robot isn't any good, then your robot's no good. But if you—it all comes back into training. The more you train with it, the better off you are with it. And you have to know what the robot's limitations are. If you don't, you're in trouble. It is a matter of … you know, again, my fallback is to training. And the limitations of how far you know you can go with it. (Simon, 49, MGySgt, USMC)

In this quote, Simon also connected the idea of robots reflecting the operator or team capabilities and limitations, too, since the current semiautonomous robots rely on human input for guidance.

Robots as Something More Than Mechanical

When recounting their ideas about how to improve EOD robots, there were variations on the idea of self-extension increasing in the technology, with robot humanlike hands to grip and move objects in a humanlike way and improved audio-visual communications to better act as the ears and eyes of the operator.

One participant, Jed, explained his idea for the perfect EOD robot as full avatar of himself. Although an outlier in terms of his detailed description of the degree of humanlikeness, his basic idea of increasing humanlikeness in robot form and functionality is not an anomaly among the other responses regarding ideal improvements in the current technology.

> *Researcher*: OK. If you could make a perfect robot for the EOD tech purposes, what features would you use or not use? Tell me about the robot you would create.
>
> *Jed*: It'd be a full human avatar.
>
> *Researcher*: A full human avatar? OK, tell me about that.
>
> *Jed*: Well, it would be me with remote control. So that I had all my capabilities, all completely into it, completely capable for everything that I could do, maybe enhanced a little bit with some kind of bionics, or something like that. But ... so that you could go completely virtual reality. Go down and do exactly what you needed to do without any kind of limitations of your own body.
>
> *Researcher*: Interesting. Would you still want to work in a team?
>
> *Jed*: Yeah, I think so 'cause a team is a lot stronger than the individual members, so if you could have two avatars down there, two, you know, two robots, which we used several times, you could always have two people get better situational awareness. They can work together, cooperative tasking, that kind of thing. (Jed, 41, SCPO, Navy)

Jed puts forth an extreme example of extending himself and humanlikeness into robot development compared to his peers in this study, but it is not discrepant from others who expressed similar desires for a robot with more humanlike affordances and abilities. Jed, a Team Leader, had previously contributed as an EOD subject matter expert representative on a military equipment review committee, and in this role had participated quite actively in pursuing new ground robotics developments.

Thus, his thinking about how to improve EOD robotics had been focused during his experiences working with the committee, and he was encouraged as part of that role to develop new ways of solving current problems with the technologies and human-robot interactions. His detailed explanation of the avatar ideas for improvement emerged from his time dedicated to considering these issues.

Although there was a clear pattern of participant insistence that robots are tools and machines, as seen in the questionnaire and interview findings, there were still a significant number of instances when participants described emotionally meaningful parts of their interactions with EOD robots. The idea of being emotionally attached to a robot was dismissed by several participants. Yet a number of significant stories emerged surrounding the possibility of attachment to robots. Emotional attachment to the robot was spoken of in three distinct ways: (1) robot as an extension or representation of [operator] self, (2) robot as mascot or zoomorphic entity, or (3) robot as humanlike other being.

During the course of the interviews, participants' facile explanation that a robot is a tool was often simultaneously couched with a portrayal of the robot as an extended version of themselves or another operator. One participant, David (22, SGT, Army), demonstrated this idea succinctly when he was asked to define a robot: "Yeah, like I said before, it's just an extension of my hands. It's a tool we use and to keep people safe."

Some interviewees went on to ascribe operator behavior to the robot, as in this exchange from Simon's interview:

Researcher: Can you tell me in your own words what a robot is?

Simon: It's a … oh, there are two definitions. One is it's a … oh, what would you say? It's a mechanical invention designed to make our lives easier and safer. The other one is … it's an extension of our own—of our own personality … . As they have to take on your—your personality after you've used them for a while. We have a tendency to think that if you have certain low attitudes that your robot, that the robot you're work—that you operate has those same things. You have a certain way you're gonna do things and that's the way that robot's gonna do it, the way you want it to do it. Well, we say it in a humorous way. You can tell the operator behind—you can tell the attitudes of the operator behind the robots by how it works. (Simon, 49, MGySgt, USMC)

Simon clarified his statement to say there is a humor component to this idea of operator personality transferred to robots, but also explained clearly how operator personalities are conveyed via the robots they use, via problem-solving choices and behaviors.

In the following example, Ben explained how his thoughts about EOD robots have evolved, and shared an anecdote about how a colleague used humor to express his feelings about a destroyed robot:

Researcher: OK, so tell me … Let's go back to your training, then, the first time that you worked with robots in The Schoolhouse. Did you have any expectations or thoughts about robots beforehand?

Ben: Not particularly, I guess. No, not really. Wasn't really something I dwelled on all that much.

Researcher: OK, and what about now? How do you feel about robots now?

Ben: I think they're a very important component of the job now. I mean, they almost become like a team member.

Researcher: I hear you say that they're important and it's ... they're almost "like a team member." Can you tell me more about that? Maybe you could give me an example of working with a robot that sticks out in your mind?

Ben: Well, you know, if ... if we had, like, personified the robot, or give it, a, you know, give it a character, or give it like a ... I mean, we would name them. And ... yeah, and if something happened to one of the robots, I mean, it wasn't obviously ... it wasn't on the same ... anywhere close to being on the same level as, like, you know, a buddy of yours getting wounded or seeing a member getting taken out or something like that. But there was still a certain loss, a sense of loss from something happening to one of your robots, and then there would be the inevitable kidding around about it like one of my friends went off ... was in Iraq, he ... an IED detonated on his robot while he was trying to do a particular operation with it and everything, and so then when they recovered the components and everything ... the carcass, if you will ... and brought it back to base, and the next day there was a sign out in front that said, you know, the guy's name and underneath of it was like, "Why did you kill me? Why?" [laughs] (Ben, 30, SSGT, Air Force)

In Ben's example, he uses words that alternately confirm anthropomorphizing robots, then downplay that significance and explain it as humor. He then uses words to refer to the robot in a zoomorphic or more detached way (e.g., the robot's "carcass").

In order to delve deeper into the territory of the first research question regarding activities, processes, and contexts that influence or constrain human-robot interactions, participants were asked about their decision-making process when a robot was in immediate danger of being harmed or destroyed.

Brady explained his emotional connection to EOD robots, and the outcome of losing a robot he worked with in close proximity for a period of time:

Researcher: Do you have any feelings or opinions about robots at this point?

Brady: They're probably the most useful tool that will save the most lives out of any tool in the Army. The sheer number of IEDs that robots have pulled apart, it's unfathomable how many lives they've saved. A good ... a team that's been through a lot is always connected to their robots.

Researcher: Can you tell me more about that?

Brady: We named ours Elly, our TALON. Yeah. And I talked to her, when I'm at the controls or trying to take something apart, caps out of explosives or whatever ... I'd be coaxing her, "C'mon, honey." [laughs] They're kind of part of the family, almost, you know? I mean, you get back from an incident, you pull a robot out of the truck, you're spraying her off, washing her off, they're all dirty or whatever. And you think about it, it's saving lives every day. So, it's very important. We like our robots.

Researcher: Were you ever in a position where the robot was in danger and you felt it affected your decision-making?

Brady: Affected my decision-making in the respect that I didn't wanna ... like I didn't want to send it and blow up the robot? Um, yes and no. We make our decisions based on ... as a Tech, we make our decisions based on how dangerous it is and how we can least put human lives in danger. There have been occasions where we didn't know what pulling on something would do. And instead of having someone put on a Bomb Suit and go down there with something with pins on it and pulling it, send the robot down. Is that going to blow up the robot? Much better than a human being. I don't think I'd really get sad in the respect that I'd miss a specific robot, because we had extra robots. But, the thing about each robot was that each robot is not the same. It has its quirks, you know, controls are looser, tighter, or whatever, and you get to know your robot. In that respect, yes, there is times that like, you know, I've had this robot for like four months now, and if it gets blown up I'll have to learn a whole new robot. (Brady, 28, SGT, Army)

Brady did name his robot and even interacted with it in some humanlike and affectionate ways, verbally coaxing it and calling it a term of endearment. However, he states that any emotional affect on his decision-making is mitigated when compared to the option of putting a human team member in harm's way. It is a choice for him between robot and human, rather than robot or robot loss. Brady also expressed the issues of operator setbacks learning the "quirks" of a new robot, and therefore his preference to keep a familiar robot when possible. This set of somewhat conflicting sentiments that sway between playful affection toward the robot and the awareness of its inorganic reality is a typical example of the RAD phenomenon found throughout the interviews.

In another example about robot loss, Jed characterized his "rush of feelings" about losing a robot during a mission:

Researcher: And how did you feel when the robot was blown up?

Jed: All kinds of things. Well, first of all, you're a little angry that, you know, somebody just blew up your robot. So you're a little pissed off about that. Just for the fact that now you're down with capability and you're one step closer to having to get out of the truck yourself. And then, you know, it's kind of like, you know, here's a robot that's given its life to save you, so it's a little melancholy, but yeah, but again, this is just a machine, a tool, that's been out there and gotten blown up, something you might have had to be exposed to, so you're pretty ... generally pretty happy just about the fact that, yeah, it was the robot and not us. So there's a whole rush of feelings going around that, and you know, the initial anger, a little pissed-offness, and just, hey, somebody blew up a robot. The fact that you've just lost a tool you've relied on a lot of times, and the fact that that tool just saved your life.

Researcher: Right.

Jed: Poor little fella. (Jed, 41, SCPO, Navy)

Jed uses anthropomorphic language here, e.g., "a robot that's given its life," and then quickly reverts to referencing robots as "tools," before referring to it as a "poor little fella." This example again illustrates the awkward accommodation managed when participants spoke about their interactions with robots. In the interviews, this sort of human language indicator was unique to referencing robots, and not used for other everyday EOD tools.

Robots were also described as companions, either zoomorphically or as an anthropomorphic other. Wade's story explains his experiences with one robot named with a traditional dog name, Fido, and another robot the operator named after himself:

Wade: I think, I don't know, I mean they all sort of took on a mascot ... Most of us named them, you know ... so ... It was the one, "Fido," and then one did "Ed." Ed, Ed. Yeah ... this guy's name was Edison, so he named his "Ed,"because of the fact you did rely on them so much, you know. They did a lot of things that up until 2003 or earlier, you know, that the bomb techs were actually still having to do on their own, so ... We do rely on them quite a bit.

Researcher: And you named yours Fido? Why was that?

Wade: Just 'cause it was like a dog. I mean, you took care of that thing as well you did your team members. And you made sure it was cleaned up and made sure all the batteries were always charged. And if you were not using it, it was tucked safely away as best could be because you knew if something happened to the robot, well then, it was your turn ... and nobody likes to think that.

Researcher: Did you just name the robot Fido, or did you paint the name on the robot or label it somehow?

Wade: No, I didn't paint it on, but it was always Fido. I'd say "Fido," and every team member knew. Ed had his name written on the arm, so it said "Ed."

Researcher: Did you name any of the other robots you worked with?

Wade: No, I didn't. I don't think so. [laughs] Just the ones you get to work with. Like I said, I think for a lot of the guys they sort of take on a mascot-type, you know, personality. (Wade, 42, SSG, Army/National Guard)

Connor shared details of the story behind one of his team's robot naming as a way to deal with loneliness via humor:

Researcher: Did you ever name any of the robots?

Connor: [laughs] Every single one.

Researcher: Can you tell me their names? Tell me more about that.

Connor: It was more just a way to be funny and keep our morale up. Towards the end of our tour we were spending more time outside the wire sleeping in our trucks than we were inside. We'd sleep inside our trucks outside the wire for a good five to six days out of the week and it was three men in the truck, you know, one laid across the front seats; the other lays across the turret. And we can't download sensitive items and leave them outside the truck. Everything has to be locked up, so our TALON was in the center aisle of our truck and our junior guy named it Danielle so he'd have a woman to cuddle with at night.

Researcher: OK, do you have any other examples like that?

Connor: Well, Danielle got blown up so obviously she needed to be replaced. I don't know … We'd name them after movie stars that we see at theater, or music artists, somebody popular, and then we'd always go to vote to decide on. (Connor, 22, SGT, Army)

From the interviews, a pattern emerged that naming the robots and assigning similar lifelike characteristics to robots was influenced by the amount of time spent with a specific robot. As in the following example from Jed, the pattern may also be influenced by the age of the operator, as well as the overall group dynamic:

Researcher: Why do you think you named some of them, and not others?

Jed: Team composition. Team in [location redacted] was a younger team, bigger team, and just couple of the more of the younger, prankier kind of guys would name 'em. Had an older, more mature team in [location redacted] and ... so it wasn't ... I don't know, it just kind of never came up.

Researcher: Did you personally, or did you notice anybody else ever treat the robot as anything other than a tool? For example, you said you named it.

Jed: Well ... yeah, it was always ... you kind of personify a little bit with the robot, anthropomorphize it, I guess. So, you know, when you talk about the robot ... and he or she, depending on which one it is. Yeah, and there's, there's actually does be ... a little bit of affection to it, especially as time goes by and ... it's done a lot of the work that could have killed or injured you, so there's a little affection drawn to it to ... It's more than just a ... you know, it's not a hammer, it's not a wrench, it's not completely inanimate. Just for the fact that, yeah, you see this out there, you see it moving around on its own or seemingly on its own doing stuff that you don't want to do. So yeah, you kind of start to lend a little humanity to it, I guess. Sort of ... not a lot, but you definitely build an affection to ... On the one side, it is an extremely capable tool that you can put a lot of reliance on, so you treat it as that. You take care of it, you maintain it, and you make sure it's capable of doing what you want it to do, and then while it's doing it, yeah, you can kind of ... you put a little humanity into it and anthropomorphize it. And I guess it just kind of helps to identify, maybe, a little with it? Or you just realize of ... how much work that's doing and, you know, you can be exposed to, so ... yeah, I dunno.

Researcher: You mentioned "he or she" depending on which one it is. What would determine a "he" versus a "she" robot?

Jed: That's the operator.

Researcher: Are you saying that they would just randomly pick a gender? Or, for example, if it was a woman operator, they might tend to call the robot a she?

Jed: No, actually, I think ... I hadn't really thought about it before, but I guess now that I think about it now, the married guys, the robots were always guys. And for the single guys, I ... which I only had two ... the robots were girls. And you know, I don't even ... I don't even know if they were talking about like ex-girlfriends or just like girls in general.

Researcher: That's interesting.

Jed: They took care of them so I guess they weren't the ex-girlfriends. [laughs]
(Jed, 41, SPCO, Navy)

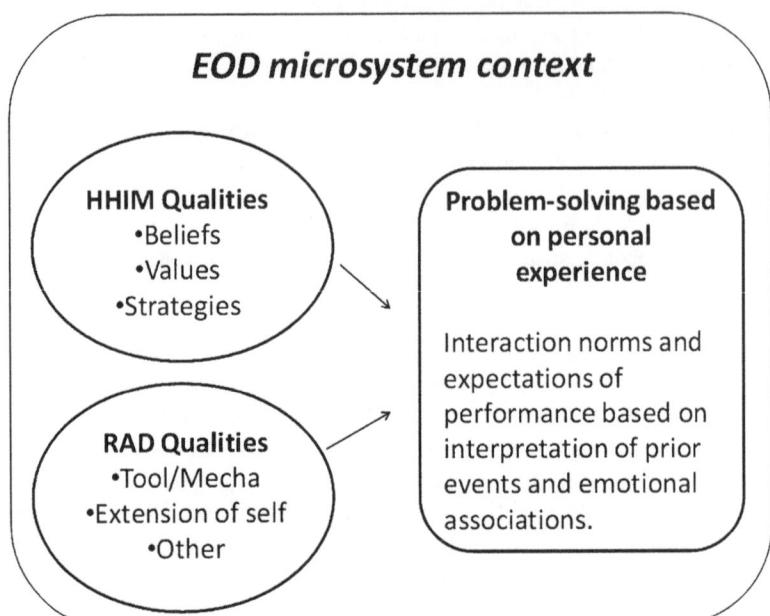

Figure 7.1   HHIM and RAD in EOD microsystem context

Jed explained that in his experience, the operator assigned the robots' gender, and like with Connor's previous example, it was an opportunity to acknowledge the human loneliness and lack of romantic companionship during deployment. Jed's explanation of the human-robot caretaking process illustrates a condition that points to some level of emotional investment with the robot, based on long-term care bestowed from human to robot. ("You take care of it, you maintain it, and you make sure it's capable of doing what you want it to do.")

In conclusion, this group of participants reported that they possessed a consistent set of beliefs, values, and strategies about human-human interactions that they practiced with their everyday group members. Additionally, they felt successful EOD work rested in this human-human interaction model they had developed via formal training and from the cultural norms of EOD as a group. The subcategory statements of *uniqueness*, *challenge*, and *family* are significant words culled from the self-descriptors of participants, indicating they have an affinity for these attributes or states.

Participants described a different set of experiences and feelings about their interactions with robots, and there was an evolving dynamic about how to treat or regard the robot consistently. The EOD personnel interviewed demonstrated an understanding and acceptance of robots as a tool or mechanical device, but also often assigned them human- or animal-like attributes. The tasks the robot performs, including being a stand-in for humans in dangerous situations, also helped inform the operators' opinions about how to categorize robots as an extension of self or tool. The danger the robot is in and chance it could be incapacitated or destroyed was reported to not affect operator decision-making. Additionally, participants' understanding of a robot's technical limitations created associated feelings of user mistrust, or at least concern about its reliability.

Moreover, the HHIM and RAD qualities are not static within the EOD microsystem of everyday work activity, and are undoubtedly affected by interactions with other social systems. For example, an individual's expectations of robot roles may be at least initially influenced by benchmark knowledge based on fictional film or literature representations of robots that have nothing directly to do with EOD work. Figure 7.1 illustrates how HHIM and RAD qualities potentially impact problem-solving.

When HHIM and RAD qualities are examined within the context of the immediate EOD microsystem of everyday group interactions, it is possible to see the potential influence on their work and mission outcomes. Coconstructed reality influences decision-making, the approach to tasks, and how goals are accomplished. This process is guided by the influence of prior experience, interaction expectations, and robot technical limitations, such as those identified in HHIM and RAD.

Chapter 8
Preparing to Repair

The primary purpose of the study described in this book was to increase knowledge about a specific population, Explosive Ordnance Disposal personnel, and their everyday interactions with robots. During the analyses of the data, it became clear that some consistency existed across individual experiences. The HHIM framework is one way of understanding and speaking about common expectations of human-human interactions within EOD work, as described by one group of individuals. Group uniqueness, needing or seeking mental and physical challenges, a high sense of self-efficacy, and practices of thoughtful analytical reflection and purposeful knowledge exchange were identified as significantly meaningful trends among their experiences.

Consistent within categories, participant relationships to one another were described using expressions such as *brotherhood, family,* and *trust.* Whether these patterns are a result of group training and general military indoctrination, personality type(s) intrinsically attracted to the nature of EOD work, or individual adaptation to the larger system of group expectations is unknown. Further research may find more common roots of these beliefs, actions, and values useful in order to better understand how organization-level policies and robot design can adapt in order to influence EOD attitudes about robots used every day.

EOD group size varies between functions and service branch, and evolves according to organizational policy and changing strategies. Each group has similar training, but individuals offer a unique perspective. Therefore, the group constructs its own dynamic. Individuals may have different motivations for joining EOD or pursuing a military career, but generally describe excitement about different aspects of their day-to-day job. The intersubjectivity of the work is often rooted in the common interests between individuals (e.g., seeking uniqueness; interest in working with explosives), and developed with the social patterns created for them (such as formal military rank), and those they negotiate and extend via an understanding of meaning within each group.

In relation to robots, the common experiences that emerged as patterns significant in meaning and through their repetition centered on experiences and concepts such as *frustration, robot as self-extension, robot as other*, and *robot as tool.*

In addition, participants described a tension between their high regard for robots as an important work tool mixed with feelings of irritation over the robots' technical limitations, and were therefore hesitant to become too reliant on robots as an ultimate solution for every mission. These conflicting reports about participant experiences with robots, and their subsequent use of the robots, are the basis for the Robot Accommodation Dilemma (RAD).

The attribution of zoomorphic or anthropomorphic traits to the robot were explained by participants as rooted in small-group/team dynamics, age of the operator, length of time working with a particular robot, troop loneliness, boredom, humor, and also self-extension of the operator's physical and/or emotional self into the robot. In other words, a variety of human-centered factors affect how operators view the level of human- or animal-like traits they assign to the robots they work with every day. Hence, the robots' design, behavior, and tasks influence these emotions and decisions, but are not the only things that affect participants' associating lifelike characteristics to these service robots. Some personnel explained the robots' evolving social role within the group comparable to that of a pet, team mascot, extension of self, or a combination of these characters. Pseudo-teammate status was also sometimes constructed on to a robot's role, but it was simultaneously understood by group members that this teammate character was assigned in a humorous context.

Intergroup dynamics may also influence interactions with robots by altering the perceived intergroup competition and perceived personal threat. People observing multiple robots may respond by self-categorizing as human, which in turn concentrates their attention on the differences between humans and robots. More machinelike or *mechanomorphic* robots may appear to look or behave even less humanlike in groups, again affecting human perception. Interconnected with this idea, a group of anthropomorphic robots may be more likely to be classified as closer to a human when there are many of them interacting in a group, as opposed to one. Thus, while robot morphology contributes to the user's social classification of these others, observing what social behaviors robots display with one another informs human social interactions with them. Similarly, other observed social activities like inter-robot competition and robot swarming behavior demonstrate coordinated actions. Human observers may infer that the robot group has a high level of cohesiveness based on these behaviors, and thus trigger an initial sense of between-group competition, human versus robot. However, any of these perceptions can be reinforced or diminished over time by interacting with the robots as individuals or a group, and therefore having their expectations supported or denied.

The core proposition of social constructivist theory suggests that social and symbolic processes produce patterns of shared concepts, understanding, and behaviors that spring from things beyond the basic acts of information processing in organizations. The findings reported here provide evidence of effects consistent with social constructivist premises, in support of existing theoretical assertions. In particular, the qualities of RAD resonate with T. Duffy and Cunningham's statement that a shared understanding of our reality is constructed "towards creating a world that makes sense to us" (1996, p. 188).

At this time in EOD work, the robot is very much described and defined by personnel as a mechanical thing—a tool. However, during the course of prolonged human-robot interaction and proximity, the robot is sometimes assigned organic traits, such as gender, association with a living person (e.g., a stand-in for the operator; or a celebrity), or inclusion in social rituals (e.g., painting a name on the robot and including it in team photos). In other words, the participants are still constructing ways of working and living with this new technology in ways that "make sense" for their social systems.

All of the study participants defined robots as a tool, mechanical system, or machine, yet many also easily assigned robots traits and characteristics of a pet or person. The people interviewed for this study were very aware that the robots were inorganic, and also generally conveyed self-awareness about any attribution they made of organicness to the robot as a playful social device, or generally communicated that they had limited emotional investment in the robots. The operators clearly stated that the robots did not merit or receive humanlike treatment, nor did they feel the robots evoked strong emotional responses purposefully (by design) or otherwise. Nass and Moon (2000, p. 20) explained similar human social interactions with computers, in which people modeled a mindful awareness of the computer as machine combined with an unawareness of treating the computers as humanlike social counterparts, as *ethopoeia*, or "a direct response to an entity as human while knowing that the entity does not warrant human treatment or attribution."

People automatically and mindlessly apply social rules to their interactions with computers because humans are inherently social. Ethopoeia, as Nass and Moon (2010) defined it in relation to human-computer interactions, has some application to this work. The study participants indicated that interactions with robots that could be identified as those based on a human-human interaction model—such as naming or otherwise assigning the robot human- or animal-like attributes—were purposeful, and done with an air of self-awareness and humor.

On some occasions, users referred to the robots in human ways, using personal pronouns and human terms, even though this study's users unanimously categorized robots as mechanical tools. Parallels between how participants described robots as people may in some cases be attributed to linguistic convenience as opposed to social considerations. Yet a pattern emerged from the data that the robots were frequently viewed as something more than mechanical. Furthermore, participants' treatment of robots was not triggered by (intentionally designed) robot social cues, although it is possible that the robots' role, design, or other characteristic unintentionally triggered a human social interaction.

Nonetheless, while interviewees may have used humanlike social rules to interact with robots in some situations and circumstances, they did not use a humanlike model to assess the robots' capabilities. Additionally, some operators indicated a sense of their own self introduced into the robot's actions, intentions, or behaviors. Thus, for the purpose of this study's findings, the term RAD may be better suited to describe this expanded set of human-robot phenomena together. The conceptual framework of RAD describes how participants are challenged by the problems of conflicting emotions, expectations, and experiences when interacting with EOD robots, and how they struggle to fix the identified problems in order to succeed. These problems are based in discovering new rules for interacting with a tool that carries out some humanlike tasks, and, in some contexts, acts as an extension of self or the operator. Belk (1988) explained that people naturally extend a sense of self into things that they control, craft, personalize, or alter (e.g., Wade shared how his team's robot operator named the robot after himself). "Objects in our possession literally can extend self, as when a tool or weapon allows us to do things of which we would otherwise be incapable," (Belk, p. 145). Groom, Takayama, Ochi, and Nass's (2009) research suggests that people are more likely to extend themselves into robots with less anthropomorphic forms than humanlike ones, indicating a possible factor in participant responses to the current models of EOD field robots.

Although there was less information about whether the social interactions with robots strongly influenced decision-making or otherwise affected mission outcomes, it is a worthy topic to investigate in further research as team configurations change and robot design evolves. There is evidence described in the findings of connections between HHIM, RAD, and problem-solving, as dynamic qualities within larger social systems (e.g., Simon explained how operator personalities could be expressed via how they choose to use their robots to carry out tasks). Therefore, this sort of firsthand information from participants indicates where to look further for points in the social systems that can be manipulated in order to change these qualities for different effects.

Human factors are a significant part of this dynamic system of interactions, and so are the part of the equation that needs to be considered in new and novel ways for new jobs that emerge when working with these new technologies. Formal work groups, such as EOD teams, are the sites of important social influences and reality construction processes. For that reason, continuing to investigate the human variables (e.g., group member age, personality, emotional effect, attachment style, team cohesiveness, and so on) of these exchanges is an important piece of understanding the overall dynamic of the human-robot interactions in any similar scenario, from the initial training stage to expert use.

This study's participants reported little initial hands-on formal training with robots in The Schoolhouse.[1] Yet the EOD personnel tasked specifically as robot operators continued training—formally and informally—once on the job. All participants described active group member roles that required ongoing communication and developing a shared understanding of each critical step in a mission, as well as post-mission analysis, in order to produce an outcome created, in part, via a social learning process. Although final decisions are made by the Team Leader, each member's contribution in the form of communication and/or negotiation is often considered an important part of the task or mission outcome (positive or negative), and because of this expectation of team behavior, every person described a sense of ownership.

---

1   The exact amount of hands-on robot training in the EOD school varies by the year the participant attended, as the formal training evolves with incoming information and new technologies become available.

Human Factors Shaping the Technology

As demonstrated with the interview stories that described jury-rigging robots in order to increase or improve their functionality, there is an immediate physical aspect to the ongoing design and function negotiation between the users and the robots. Another example of the iterative design and function negotiation is direct user feedback to the robot developers. One example of direct feedback to robot-design decision makers was Jed's service on the military equipment review committee, where he had the opportunity to provide suggestions about the directions robot design might take. In both jury-rigging and design feedback examples, there is an obvious circle created as the robots (or objects) impact EOD, and then EOD create and influence (construct) new ways of working with and designing robots.

One intent for this study is that it would lay the groundwork for investigations into related research. Therefore, the following questions reflect a synthesis of the data presented, and the discussion of those findings.

To summarize the findings:

1. There is an identifiable human-human model of interaction within the studied group with clear expectations, beliefs, values, and strategies.
2. Operators categorize EOD robots as tools, but sometimes interact with them in ways that resemble human-human or human-animal social interactions. Participants used linguistic indicators of inclusive familiarity, category membership, and perceptual similarity to sentient beings discussing robots. At the time of this study, there was no indication that participants attributed internal mental states to the robots.
3. A separate interaction model with its own parameters and expectations exists between user and robot, forming the RAD dynamic.
4. Unlike HHIM, RAD is a one-way social model since (EOD) robots are not capable of returning purposeful social signals or communication. Furthermore, robots' lack of reciprocity and inability of the robot to be a fully participatory social actor is perhaps a significant parameter of limiting them to the users' "tool" category at this time.

Robots are a technology that transform personal experience and social relations by forcing users to find new ways of acting with this new sort of agent; the robot is a tool, but one that performs some humanlike functions and actions.

Questions that arise from this study include:

1. What organizational factors are producing social-role changes in EOD robot use and are influencing the social dynamics between EOD and robots used every day? Examples of these factors include (a) standardized Schoolhouse (e.g., training) procedures, (b) evolving (robot) designs/behaviors/roles,

     (c) changing (EOD) group size, (d) popular culture representations of robots, and (e) a combination of these inputs.

2. What impact does everyday robot use have on team dynamics? What are the outcomes of these changes?

3. What are the different human-robot interaction patterns for EOD in each branch of the military?

4. How can researchers deeply explore other military subgroups that use robots every day, such as those working with UAVs, UGVs, and similar unmanned semiautonomous systems?

5. Why is a robot viewed as an extension of self? What frustrations with the robot's capabilities are connected to the operator's sense of frustration at self? What sense of self is lost or lessened with the unintentional loss of a robot, and does that influence user emotions and behaviors?

6. What level of inclusiveness will be most effective for human-robot teams in different defense collaboration situations, such as in EOD work or for RPA operators?

7. How can roboticists leverage any human tendencies of projecting a sense of self into the robot design, behaviors, and tasks?

8. Which robot physical appearance, behaviors, and tasks trigger human tendency to anthropomorphize or zoomorphize robots? When are these triggers desirable for military scenarios and when should they be minimized or eliminated?

9. What identifiable existing robot affordances are examples of triggers for human sentiment, preference, or affection? What are the effects of human sentiment in human-robot teams, especially in militarized (and other stressful) settings?

10. What conditions, uses, and design cues leverage or diminish human evaluative judgments about a robot as an *individual*, unique from the larger category of "robots"?

11. What trust and team cohesiveness human factors will arise as robot physical appearance, behaviors, and tasks evolve? What do human-robot trust models look like, and does human-robot trust develop in a model similar to human-human trust?

12. What level of responsibility will operators feel as robots take on increasingly complex and autonomous tasks? How will these changes affect team members with a high sense of self-efficacy or achievement?

13. What social, cultural, and psychological outcomes occur when humans interact with multiple robots in one setting? How will variables like group-coordinated robot behavior, or teams of human-(multiple) robots working toward common goals, affect these interactions and processes?

14. What (new or existing) stress issues surrounding human-robot interactions arise if users extend their selves into robots used every day, or otherwise imbue them with human- or animal-like socialness?

Based on the areas of future research suggested here, there is a need to develop rich psychological scales for measuring the mental states of robot users and analyzing related social trends over time. In particular, the body of existing literature (Bates, 2002; Carpenter, 2013; Hogan and Hogan, 1989; Kolb, 2012; Mori, 1970/2012; Murphy, 2004; Scholtz, Young, Drury, and Yanco, 2004; Singer, 2009) supports the need to explore the human factors related to human-robot interaction, and to attend to the issues related to operator stress and anxiety.

Results in this work also hint at user variables such as operator age or group dynamics affecting the tendency to humanize the robots. Therefore, further research into the human side of the equation must continue in order to better design robots that most effectively work with humans on tasks.

Revealing Paradigms

After this study was initially published as a dissertation, there was further feedback from people who felt this area of research resonated with their own experiences. One touching e-mail was from a person who quite generously shared this bit of his story (Hay, K., personal communication, October 2, 2013):

> I recently read a small article about your research into EOD personnel and their attachments to their robotic platforms. As I am an EOD technician of eight years and three deployments, I can tell you that I found your research extremely interesting. I can completely agree with the other techs that you interviewed in saying that the robots are tools and as such I will send them into any situation regardless of the possible danger.
>
> However, during a mission in Iraq in 2006, I lost a robot that I had named "Stacy 4" (after my wife who is an EOD tech as well). She was an excellent robot that never gave me any issues, always performing flawlessly. Stacy 4 was completely destroyed and I was only able to recover very small pieces of the chassis. Immediately following the blast that destroyed Stacy 4, I can still remember the feeling of anger, and lots of it. "My beautiful robot was killed … " was actually the statement I made to my team leader. After the mission was complete and I had recovered as much of the robot as I could, I cried at the loss of her. I felt as if I had lost a dear family member. I called my wife that night and told her about it too. I know it sounds dumb but I still hate thinking about it. I know that the robots we use are just machines and I would make the same decisions again, even knowing the outcome.
>
> I value human life. I value the relationships I have with real people. But I can tell you that I sure do miss Stacy 4, she was a good robot.

However, not every response to this work has been this in-line with some of the more distinct patterns of the findings from the study. It must be acknowledged that there exists that squishiness of ideas that can be the root of frustration when discussing human emotions. Information of this kind is challenging to interpret and act on, especially in a timely manner across expansive organizations. Once society entertains the complexities of human preferences, expectations, and feelings about technology and their impact on everyday interactions, it is no wonder people become overwhelmed with how to interpret and apply these fuzzy things in everyday life.

After the initial study, one more interview opportunity presented itself, another one-on-one conversation. Although this conversation was not included with the original 23 EOD stories, listening to an expert is always valuable and adds to richness an existing body of knowledge. The following is an excerpt from that informal exchange about EOD work, and one person's personal experiences with EOD robots in the Marine Corps.

> *MSgt R*: The robot is just an extension of oneself. You can make that robot do what you need it to do … . It's all based on how well you operate it, and how you practice and train to do that mission. A robot will never replace the operator. You have to have someone there. Someone has to have the requisite skill sets to understand what is down there. If the robot fails, you have to go. … Folks get attached to the robots. It is a team member. Here's an example. Our first robot, when it was blown up, we had a helicopter come in and it was treated just like a casualty. It was picked up by one bird, it was flown all the way to Baghdad, and then we had it back in four days. The robots were a very critical piece for mobility and survivability out there. … We never had the attachment or the time to do burial ceremonies or anything to our robots. Typically, if they were blown up, we would gather what pieces or parts we could, wrap it, and leave clear the area. We get back to base, everyone vents their frustration because, "Hey, that was a good robot." We had other robot riddled with explosions, and we held them together with duct tape, because they kept working and they were reliable until the electronics suffer and just let go. And then, OK, it's gone. Get a new one.

> *Researcher*: Did you ever name any of the robots?

> *MSgt R*: No. Probably just because I was anal. I'd say, "Get the crap and let's go." Make sure it's all loaded up … We kept the robots in the building some of the time, but the majority of the time we kept them in the back of the truck. So we didn't see them until we cleaned them, we prepped them, and we treated them like weapons. As long as it's clean and you take care of it, it's gonna be there to function and do what it's supposed to do. But their place was in the back of the trucks, secured down unless we had to pull them out for maintenance.
>
> (Roberson, S., personal communication, March 5, 2014).

In this passage, the possibility of personal attachment to the robot as other is acknowledged, and explained by Roberson in several ways. There is the clear, unprompted identification of the robot as an extension of self. Moreover, robots were part of the team's shared experiences, including the everyday caretaking of the robot, and the end emotional bookmark to the relationship by the team's collective expression of frustration at its loss. Yet, he also clarifies the robot's "place was in the back of the truck," and that they treated them "like weapons."

Ultimately, these stories are characterizations about what is significant from a specific point of view. In due course, all cultural shifts of opinion and actions, even those of war, are built on standpoints that become collectively agreed upon activities, but on a much grander scale.

Chapter 9

Transformational Shifts

In these early stages in the field of human-robot interaction studies, issues of accurate user expectations of robot functionality, behavior, and responses are growing increasingly more complex and raising questions of ethics and morality in relation to human-robot relationships. How robots will be used in warfare and covert operations are a set of ethical and political debates that will be ongoing. Growing discoveries in robotics, human emotional reactions to new technologies, and new situations and contexts for robot use create new human-robot relationships that lead to larger discussions relating to the expectations, obligations, and responsibilities humans have toward machines and their uses (Brooks, 2002; Lin, Bekey, and Abney, 2008; Arkin, 2009). Therefore, investigating individual expectations of robots is not merely the domain of the individual user, but also of society. The world is, essentially, merely at the beginning of inquiry and theory building in all aspects of human-robot interaction as a field of research.

Discourse about human-robot ethics and policy has already emerged in the form of such initiatives as South Korea's *Robot Ethics Charter*, which claims to be developing legal guidelines on how to treat robots (Yoon-Mi, 2007). In addition, there are projects like the *Euron Roboethics Roadmap* (Veruggio, 2006), developed by scientists from the European robotics community who are responding to the need for discussion and development of an ethical framework that may eventually serve as a useful guideline for the design, manufacturing, and use of robots. In the United States, experts have called for a Federal Robotics Commission (FRC) to be established. The argument for forming this consortium is that robotics—and our human interactions with robots—are unique situations that challenge existing laws and policies in place for other technologies.

Founded in 2009, the International Committee for Robot Arms Control (ICRAC) is an international committee serving as a non-governmental organization (NGO) "concerned about the pressing dangers that military robots pose to peace and international security and to civilians in war" (ICRAC, N.D.). In 2014, ICRAC's mission statement outlined their position that machines should not be in the position of using autonomous decision-making for violent force or killing. This statement also calls for:

- Limitations, regulation, and transparency for remotely controlled unmanned systems.
- A ban on new kinds of unmanned nuclear weapons delivery systems.
- A prohibition on robotic space weapons.

ICRAC is also a member of the Steering Committee for the NGO *Campaign to Stop Killer Robots*, which urges social responsibility through activism focused on banning autonomous weapons in general, although the name of the group specifically emphasizes robots. Lethal Autonomous Robots (LARS) differ from drones and similar weapons operated at a distance because without human interaction, they are capable of determining when to strike a target. In a report to the United Nations Human Rights Council, Heyns stated, "War without reflection is mechanical slaughter. In the same way that the taking of any human life deserves as a minimum some deliberation, a decision to allow machines to be deployed to kill human beings deserves a collective pause worldwide. ... " (Human Rights Council 23rd session, 2013).

The objections to autonomous weaponized robots span technical and ethical reasons. Within a few decades, perhaps sooner, robotic weapons will likely be able to pick and attack targets—including humans—with no human controller needed. Steven Goose, a supervisor of arms-control activities at Human Rights Watch, a group that manages the efforts of approximately 50 non-government organizations supporting the ban, stated, "We're not anti-robotics and not even anti-autonomy. We just say that you have to draw a line when you no longer have meaningful human control over the key combat decisions of targeting and attacking ... . Once you reach a stage where a platform is weaponized, you have to maintain meaningful human control." Based on a long history of human-centered warfare, many of the existing rules of military engagement and the nature of war would be difficult for lethal autonomous weapons systems to navigate. Subjective concepts such as necessity, the ability to distinguish enemies from comrades, and the complicated considerations when weighing the potential harm to civilians against military advantage in any action are all currently beyond the capabilities of these systems.

To opponents of these autonomous weapons systems, the fact that these robotic systems cannot meet the requirements of International Humanitarian Law and International Human Rights Law needs careful consideration when developing policy. Additionally, as a matter of ethical concern, there is the position that robots should not be given the power to make life-and-death decisions about humans.

Yet supporters for the advancement of research and development of autonomous weaponized systems maintain that these will be advantageous tools of war because of this very lack of humanness in the machines. Advocates for the systems claim another danger emerges from a total ban: that of the limitation of scientific advancement of technology that could save human lives. Their argument is that while autonomous weapons lack empathy, the closed systems are also not influenced by human emotions such as anger, fear, or revenge, and are not subject to human physical conditions like fatigue or hunger. Additionally, eventually this technology will make faster decisions, and have a more responsive capability in rapidly changing conditions than humans. Proponents of this line of thinking argue that to reject the development of this technology is irresponsible, as the weapons will become produced and available in the world regardless of formal policy.

How people in one group perceive others viewed as outside that group, and what criteria are used to make determinations between people in groups in order to justify treating others as less than human are important distinctions in war, and in how people think of themselves. It is widely acknowledged that there is a dehumanization process of *other* people, those who are regarded as part of the opposition, during warfare. Making these others something less than human or dangerously different from what is recognized as similar to the "home" group is one way to gird individuals emotionally for the violence of conflict, and the inevitable loss of life. How people develop these distinctions between *them* and *us* is, in many ways, similar to how they develop social categories and distinguish themselves from nonhumans. What is still being discovered are *how* and *when* a sense of self becomes entwined with an other, like a robot, and if these things change the definitions of what it means to be human. The moment when people come to feel an affection for an object, and attribute attachment, sentiment, history, and a uniqueness to it, is when they impart a quality that is, in some ways, humanity.

The Nature of Drones in War

At one time, the very word *drone* denoted these things: monotonous activity, an incessant sound, or a thoughtless worker that follows orders. Unmanned or Unpiloted Aerial Vehicles (UAV), also now referred to as Remotely Piloted Aircraft (RPA), have quickly become perhaps the primary associations for the word. Perhaps unfairly, the allusions of past meanings of the word *drone* linger in our language, for now.

Although team roles are changing rapidly as technology changes, a representative operator scenario for an RPA operation consists of at least two people who work directly with the drone, a sensor and a pilot. Other people, such as imagery analysts and safety analysts, interact with the robot consistently but indirectly to monitor its actions and their repercussions. The sensor and pilot sometimes share the same physical space, seated near each other. The pilot controls flight maneuvers while the sensor is responsible for manipulating the robot's cameras and aiming its weapons at selected targets. The process of weapon launch requires cooperation between these two people who initiate and direct the drone's actions, or behaviors. When the pilot fires, the sensor directs the missile to its ultimate target. The analysts may watch the video feeds of the drone missions live, or be required to view them repeatedly in order to verify details.

According to the accounts of some former drone operators, it is purportedly feasible for these teams to be assigned to work twelve-hour shifts, occasionally overnight, up to six days a week. A Predator drone can stay airborne for 18 hours, and pilots-sensor teams can be expected to perform in tandem with the endurance of the technology.

Operators also report that as part of their job, they observe people engaged in their everyday lives. These people—referred to as *targets* in military jargon—are sometimes unaware of the drone eyes projecting their image around the world to the sensor and pilot. One former operator said if his mission was to monitor a particularly important target, he might remain above a single house for weeks.

One operator who spoke publicly to news sources about the emotional processes he went through found himself one of the first voices to tell his story to the media, sometimes to negative feedback from comrades or others who believed his emotional experience was less real than those who had a boots-on-the-ground involvement in war. As a part of a drone team, the first time he fired a missile, it killed two people and mortally wounded a third while he watched their prolonged suffering on camera. His response was to cry on the way home from work, and then call his mother for comfort. He also claimed an increasing sense of "disconnection" from humanity, from others (Abé, 2012). Assigned to gather intelligence and perform reconnaissance on known Taliban fighters, he saw other people identified as worthy of surveilling go about their daily lives: children playing, farmers working in their fields, husbands and wives embracing with affection. "I got to know them," he explained. At night, switching to infrared technology, he even witnessed moments of intimacy between couples, "two infrared spots becoming one" (Abé, 2012).

There is, of course, also the trauma of watching comrades die. Image analysts view reconnaissance and weapons launch and impact day after day, observing casualties and kills while unable to physically intercede, and afterward, living with what is perhaps a version of survivor guilt. Moreover, sometimes there can be questions about how accurate their own identification of hostile forces or weapons confirmations were and the choices they made based on these observations. In all aspects of warfare, accountability is a tenuous concept. RPA operators may be physically located half a world away from the location of a mission or incident, but they take still very much take a significant part through their actions.

There is little data available on RPA personnel mental-health issues related to or exacerbated by their work environment. Self-reports from a handful of former operators indicate that this new way of self-extension into war via great distance can be significantly traumatizing. While a sensor may not fire the weapon, the coordination effort is still ultimately part of deadly force. RPA operators may not be at risk for physical injury in battle, but the weight of their actions affects mood, sleep, and can contribute to thoughts of suicide and clinical depression. Unlike the popular persona of "coffee cup warriors" that some have dubbed RPA operators, the reality is that they go to work for many hours and many missions, and take part in operations with violent outcomes. Then, they go home to their families at the end of a work shift, and, at some level, need to shut off and compartmentalize their emotions. During a work shift, their actions have unalterable consequences for others, and the "switching off" of a natural emotional process to go home and live with that understanding is a different kind of trauma than what has been seen before in war.

In a 2011 Air Force mental-health survey of 600 combat drone operators, the report authors discovered that many operators felt a sense of accomplishment by aiding and protecting troops on the ground (Chapelle, Salinas, and McDonald). Yet this study also reported that 40 percent of Remotely Piloted Aircraft (RPA) Predator/Reaper crews reported moderate to high levels of stress. More specifically, self-reported rating of occupational stress revealed that 15.3 percent of the participants felt very or extremely stressed, while 19.5 percent reported high levels of emotional exhaustion. These findings are in contrast to a control group of 600 Air Force members situated in logistics or support jobs, where 36 percent reported stress. There were no comparisons between RPA operators and military pilots who pilot planes.

The position of the RPA operators is understandably different from that of others who are part of the same interaction, on the ground. The soldiers on the ground who see the drone may view it as a protective or otherwise reassuring presence. Civilians within theaters of operations may view the drone with a mixture of emotions, from fear of the unpredictability of war as a whole, to indignation at the presence of weapons in the skies, to a sense of comfort that allied forces are in the area. Enemy forces, of course, view the distinct presence of drones from a completely different place emotionally, likely running a gamut of anger, frustration, or even humiliation.

While the means of killing may change over time, whether using a bullet or a laser, the finality of the deaths remains the same. Debates will continue to center on whether or not drone operators can experience trauma in a way that falls under current formal psychological categories. How to quantify or categorize the toll it takes on the human psyche to kill from a great distance will formally evolve. Pinpointing different aspects of the situational differences and their effects on emotional trauma will shed light on the differences in physical situations. Still, the lingering and pervasive feelings of terror and guilt have different causes, but the challenges remain.

Another relevant concern about the interactions from the operator viewpoint is the possible extension of self to the drone, or aircraft. Whether this self-extension occurs, to what degree, or with any regularity, remains to be proven. Similar to the interface design of some EOD robots, there is a purposeful integration of video game-like aspects to operating a drone, like the use of a joystick. While this is a useful basis of familiarity for many users, it also draws research parallels to existing ideas about player extension into game characters.

Drones are not going away any time soon, although their design and uses will evolve. One suggestion is from the University of Washington's Ryan Calo, who has proposed increasing the anthropormorphication of drones in an attempt to alleviate pilot/operator guilt. By giving the drones more humanlike characteristics, such as natural language ability for basic interaction, the theory is, in part, that this type of interface will disburse the guilt related to operator decision-making. The hope is that the RPA operators will order the drone to act, and will feel that it was less their direct action because the ultimate responsibility is shared with this anthropomorphized other (2012, Axe).

Self and the Machine

There are extraordinarily different cultural variations of how the ideas of *self* and *other* are constructed, and of the interdependence of these concepts. In turn, these cultural variations are factors that influence individual experiences, including cognition and emotions. As an example, in American culture, one meaningful way people seek to establish self and independence from others is by expressing their unique inner qualities. This view of an individual is as an independent and autonomous entity with unique attributes, whose motivations and behaviors are believed to be based on these distinctive traits, abilities, motives, and values. As another example, in some cultures there is a view of self-regard as something that is interdependent and reliant upon surrounding context, where self is discovered and defined in relation to others. In this way, *self* and *other* are ideas connected to everyday tasks, things that people are expected to do as part of their lives.

Furthermore, beyond a physical or ecological sense of self, each person has an awareness of internal activity, such as the ongoing activities of thoughts and feelings, which are private to the degree that others cannot directly know them. The awareness of this private, unshared experience leads to a sense of an inner, private self. The exact content and structure of the inner self may differ considerably by culture. It also presents the foil for the *outer* or *public self*, which is derived from interactions and relationships with other people. Like the other frameworks of self, the nature of the outer or public self that derives from one's relations with other people and social institutions may also vary significantly in different cultures. Important aspects of understanding the functions of self and other are their roles in motivating people, or otherwise inspiring them to action.

Things that make up some of the fundamentals of self are:

- Personal physicality, e.g., facial features, organs, body.
- Psychological processes, e.g., the conscience, emotions, sexuality.
- Identifying characteristics, e.g., name, age, career, or job.
- Possessions, e.g., clothes, house, vehicle.
- Production, e.g., output of work, actions.

Physical proximity is a determinant of an item's inclusion in the definition of self, and objects over which a person has control or that can be manipulated are more likely to be classified as part of the self. In this way, other people such as family, friends, and coworkers, objects regarded as unconscious, such as furniture, and abstract ideas, like social norms or laws, are not typically regarded as part of the self. This theory of self provides a framework for understanding the human-robot relationship from a position of social presence.

In computer-mediated environments, *presence* occurs when interactions with people, avatars, and objects feel or appear real, direct, and immediate. Biocca, Harms, and Burgoon (2003) highlight that presence typically consists of two interrelated phenomena: (1) *telepresence*, or a sense of physically being there, and (2) *social presence*, the sense of being with others. These two forms of presence are evident in virtual and computer-mediated worlds in which people negotiate telepresence, or their degree of immersion and engagement in the space, through their interaction with their avatar, and social presence through their interaction with others as an avatar. It is perhaps not surprising then that a virtual environment that offers some of the same criteria for self-building in the real world can encourage identifying the self with an avatar or other distance representation of their physicality. As outlined in Schultze and Leahy (2009), a virtual world may have:

- Customizable avatars that enable computer-mediated embodiment.
- In-world virtual possessions such as weapons, currency, furniture, or clothes.
- Animations within the avatar and its environment that allow movement and interaction within the virtual environment.
- Information such as name, interests, and group memberships or affiliations in a profile shared with others.
- Cameras that facilitate vision and gaze from various viewpoints.
- Means of communication such as voice, private or public chats, and note cards that can act as documentation.

These ways of being in the world can support or limit a person's sense of existing in another environment, whether in virtual fictional surroundings or a representation of a real environment at a distance, such as in teleoperation contexts. Combined, these factors—facsimiles of ways people interact with their environment in the real world—enable emotional and cognitive immersion both through interaction with their avatar or virtual self (telepresence), and via interaction with others as an avatar (social presence).

Given the significant contribution that others have on constructing individual ideas of self and the associated version(s) of reality and methods of regulating individual and group behaviors, understanding self and otherness is worthy of investigation in order to discover which, if any, parts of these processes are universal. Moreover, to provide important understanding into subcultural variation of these processes, an examination of groups such as the military—or even more granular, specific groups, such as EOD—has potential for insights into controlling aspects of these activities. Encouraging or mitigating a sense of self or otherness in the real world allows or restrains individual factors such as motivations, behaviors, and engagement, as well as group interactions like collaboration, trust, and cohesiveness. A theoretical understanding of these phenomena leads to successful practice of modifying or regulating human behaviors in virtual and real environments, working with people, robots, or other rich technologies. Coexisting with others in any environment requires people to continually assess and reposition their own actions, depending on the interpretation of other as a danger, as supportive, or even as an entity that will improve the self by association.

The significance of others in constructed reality and the distinctions made between threatening and supportive others is vital for the interdependent states people live in. The boundaries of a social group are dependent on the concepts of self and other. For cultures where the representation of self is developed around the ideas of uniqueness or distinctive from others, identifying others becomes about determining if the other is like the self, rather than questioning whether the self is similar to the other.

People often have an erroneous tendency to perceive behavior as an outcome of internal characteristics and processes of another person, attributing actions primarily to the other's personality, rather than the situation or context that also affect behavior (Ross, 1977). Now, people use this attribution effect to interpret robot behavior, at least as an initial or latent mode of understanding. The degree to which this bias is applied is affected by factors such as cultural influences, operating within an environment of high cognitive load, or individual tendencies such as people with a high locus of control.

Becoming attached to an avatar as an extension of self or as a representation of self is a legitimate form of self-concept. When attachment occurs, harm or the threat of harm to that representation may register as a moral or ethical wrong to the avatar operator. To say that this sort of perceived threat to self is insignificant because it is "not real" minimizes the attachments already believed to be significant, such as attachment to people, pets, and possessions. In a virtual environment, the avatar is the operator's representation used to communicate and interact with the world. Even if the virtual representation of the user does not have physical characteristics similar to the operator, the avatar actions are an extension of the controller's decision-making, personality, and belief system.

Unfortunately, understanding attachment to a virtual representation of self does not provide a prescription or rule book of how to use this information, or a means to predict what issues and challenges will arise as other situational factors vary. For example, if an operator is emotionally attached to an avatar, does this hinder or help them work effectively through it toward their goals? One clear concern in defense or other dangerous and stressful work is that a heavily emotionally invested operator may become distracted upon the loss or harm of an avatar, whether it is a robot or other complex computer-supported representation of the soldier used for engagement via distance.

One option to alleviate this particular issue of avatar harm could be to encourage users to stay emotionally remote from their avatars, such as through training aligned specifically on this topic. Yet this solution assumes that operator attachment to this representation is not morally significant, and places the onus of emotional detachment on the operator, an instinct that would be difficult to control reliably across time and various situational contexts, even mindfully and with extensive training. When using robots as avatars, such as teleoperated semiautonomous robots, the idea of explicitly telling operators to emotionally divest themselves is a tricky prospect. As robots become more commonly integrated into people's everyday lives in many situations, operators' fluidly changing emotional expectations of interaction with semiautonomous robots that are deemed worthy of emotional investment versus those from which one must reserve emotions will be—at the very least—an ongoing cultural and cognitive adjustment.

Interacting with robots at a distance presents highly situated and dynamic interactions for the human operators. Still, as with other interactions with rich technologies, it is possible for an operator to engage with the robot in a dynamic fluidity. An operator may seamlessly fluctuate from a relationship with the robot-as-avatar, perceiving it as merely a tool manipulated to work toward a goal, and then shift to perceiving the robot as an extension of self because it enables the operator to become immersed in a distant environment. Moreover, as robot morphologies, customization options, and situational contexts vary, an operator may regard a robot avatar as an idealized representation of self. For instance, customizing a teleoperated robot to appear especially fearsome in a war situation may bolster the operator's assertiveness, which could lead to either helpful or overconfident or risky behaviors. Additionally, an operator may use a teleoperated robot as a vehicle to express actions they would not normally demonstrate as their human selves because there is a sense of anonymity—a lack of personal recognition of the operator by those targeted—when controlling robots at a distance.

As seen with the small group of Explosive Ordnance Personnel interviewed here, very different individual views of robots do not mean the phenomenon of robot attachment does not exist. Rather, there is a system of models of thinking about the robot that are contextually and situationally dependent. The boundaries used by individuals within a culture to define identity and self are socially constructed and fluid. Manipulating robot design and use based on what is known about these things may not lessen people's tendencies to become emotionally attached or to self-extend into the robot. Yet each bit of gained insight into the human side of human-robot interactions contributes to an understanding and influence of controllable variables (e.g., design, training) for the most positive intended outcomes of human-robot teamwork.

Robots, even very humanlike ones, are not something neatly (or regularly) categorized as another person in an environment at this time. Still, how people position themselves in relation to robots informs their connections with other robots they encounter, and subsequently informs how they might interact with them.

Cultural Evolution

People have always learned new ways of existing with technology as media for connection with other people. From writing a letter to sending e-mails, whether using a telegraph or a smartphone, society has navigated new customs and expectations of the capabilities and limitations of these things to communicate with people. One way to examine the interrelated and dynamic circle of influence between technology and culture is to take the position that western models of the self are at odds with actual social behaviors. Eventually, inevitably, these paradigms will evolve to reflect the substantial interdependence that characterizes even western individualists. The reality of globalization forces a rethinking of the nature of the individual.

Another example of extending self is through our use of social media. Social media has provided rich means for extending our presence, while sometimes simultaneously making the world feel a bit smaller when someone can reach out to another person at a physical distance and share experiences. Facebook and Twitter are two currently popular and even pervasive examples of how people interact through a social technology, albeit computers are viewed as a disembodied entity. An interesting occurrence in recent years has been the adoption of accounts on sites like these as a vehicle for pseudo self-expression from the point of view of a technology, sometimes robots. Although managed by real people, the narrative structure of the messages and interactions with the readership are based in a human-centered social model of exchange. These accounts are generally framed as transparent attempts to engage people, whether the intent is humor-based or to gain interest in a particular event or project.

One example is the Twitter account set up by Isa Barbarisi, a European Space Agency engineer. LVX-1 (@isa_MYB), voiced by her in a narrative worded to resemble a robotic tone, presents transformed images of humanlike robots in space, captioned with LVX-1's musings about space travel. The name LVX-1 is taken from an Isaac Asimov short story, Robot Dreams (1986). The intertwined worlds of storytelling—from science fiction and an unfolding story centered on real space missions—converge in one account. "I work for ESA ExoMars mission. This account is to inspire people and bring them close to these exciting space missions," Barbarisi explained (personal communication, January 12, 2015).

In 2013, at the International Space Station Kibo Japan Test Building, Kirobo, the diminutively sized humanlike robot, engaged in conversations with astronaut Koichi Wakata. Many around the world watched videos of the two speaking in a natural way. In fact, Tomotaka Takahashi, the robot's designer, explained the overarching goals for developing Kibo (Burton, 2014):

> The Kibo Robot project has a special mission. To help solve the problems brought about by a society that has become more individualized and less communicative. Nowadays, more and more people are living alone. It's not just the elderly—with today's changing lifestyles, it's people of all ages. With a new style of robot-human interface, perhaps a way to solve this problem can be found. This is the goal we have in mind for this project.

The project mission is a robust one, and it is difficult not to feel empathy for the weight of the goals on the small robot's shoulders. Perhaps the most poignant moment between the two was the final exchange between Kirobo and Wakata, when the robot asks the astronaut not to feel bad about saying good-bye. "I'm a little tired, so I think I'll rest a while, but I hope you'll look up at the sky sometimes and think of me," (Burton, 2014).

A military satire Website ran a story titled *EOD Soldier Files Discrimination Suit After Army Denies Marriage to His Robot* (Duffelbag, 2013). Meanwhile, nonfictional firsthand stories of troops customizing EOD robots for humorous ends circulate on the Web. The robot developers sometimes engage in a type of fun and affectionate display for the product of their work, such as when a very real wedding between a QinetiQ mechanical engineer and her fiancé was attended by a DragonRunner robot acting as ring bearer. The robot entered the ceremony to the music of Styx's "Mr. Roboto," was outfitted in a small tuxedo, and joined the partygoers on the dance floor. From all reports, this wedding was a successful celebration of love, including the robot's presence and inclusion as a meaningful reference to the bride's everyday engineering work (Pantozzi, 2012). These examples exemplify part of humanity's negotiation with these mechanical things that are built, worked with, maintained, and often depended upon to be effective. These interactions are illustrative of the evolving relationship with robots as sometimes-social machines.

Perceptions of truth are dependent on agreed-upon conventions, human perceptions, and social experiences. The trueness of something can be evaluated, in some ways, by whether or not it follows coherent and consistent concepts across a system. Truth, then, becomes an element of a larger whole, where individual factors are considered part of the greater system. Yet the rules of logic do not apply to the dynamics of what is understood as the truth for each person.

There can be searched-for patterns to establish a coherent series of facts, and that is an effective strategy for understanding phenomena, to a point. There are consistencies in versions of truth, and those are perhaps more easily probed and pulled apart for analytical discussions. Still, all of the experiences people have cannot so easily be collected, sorted, and examined. Some do not consider emotions a test for what is true in the sense that gathering evidence to support human instincts, memories, experiences, and intuition is an impossible task, but observable things are also not the only way to determine truth for each of person.

A larger narrative is emerging, one about humans and robots and how it is not only acceptable, but also possible to integrate them into the world and people's everyday lives. We are writing a new story every day, all of us, together.

Appendix A
Verbatim Participant Definitions of a "Robot"

Name	Questionnaire	Interview
Aaron	A machine with multiple functions operating either autonomously or under remote control, not just a vehicle.	It's basically a machine that can perform multiple functions, you know, not just a single function device either autonomously on its own or when its programmed to go do something or with direct control of a person who's not sitting there physically touching it, with either a radio controller or whatnot. Where you're sitting there with a controller and it's off somewhere else doing its thing.
Brady	A complex tool used by humans to achieve required results.	In simple words, it's a tool. A very, very important tool. It's complex and there's wires and circuits and cameras and all that, but when it comes down to it, it's just a tool. The tool we use the most, and very expensive, but it's a tool.
Irving	Any system which through direct human control, semi-autonomous, or fully autonomous function, perform a function, service or action through electromechanical movement.	Something that performs through electro-mechanical function, um, performs some type of movement, some type of function, be it … I'm gonna quote myself a little … either through direct human control, autonomous or semiautonomous control basically to either mimic a human behavior or to perform a designed action.
Sarah	A remote tool.	I said just basically it was a remote tool … to be able to render a bomb safe or to dispose of it with remote capabilities. Keeping, you know, all persons at as much distance as possible, as its capabilities will allow.
Jeremy	For EOD purposes, it's a tool we use to control remotely to recon, video, manipulate, place tools, etc., on IEDs or suspicious items in lieu of sending a bomb technician down to investigate.	It's a … .for EOD purposes … .it's a tool we use to perform remote reconnaissance, manipulation of devices, to investigate unknown items, and to place tools, demo charges, using that, from a remote distance.

continued …

Name	Questionnaire	Interview
Hector	Motorized mechanical human controlled or programmed tool.	I guess there's a lot of different types of robots. It's a machine. It doesn't make its own decisions. It's either directly controlled or it could be programmed to do a certain job. Usually they move about in some form but not all of 'em. I don't know. I was a … I used to work at [redacted] in [redacted], and we had robotic painters. And so those, they moved an arm, but they didn't roll around or anything. So they were robots and nobody controlled 'em directly but they had a computer program that controlled 'em. So I consider those robots, too. I guess that's kinda hard to put into words.
Marshall	A tool to complete a mission which prevents me from taking a risk myself.	I really think of the robot as a tool and an extension that I could-, it's a tool that allows me to do my job, and not take particular risks. OK? I mean, my robot, it was very-, very important to me. My most important tool because it gives me the thing I need the most when I operate and that is distance. I mean, if I can disarm that bomb or at least figure out, even if I can't get to it or disarm it with the robot, I have much better situational aware-, awareness after running it down, seeing it and then when I've put on the bomb suit, I can … I can get down there and I'm much less likely to get killed if I know what's down on the ground before I actually have to go see it with my own eyes. The ro-, it's an extension of yourself that when you become a good operator with a robot, it's … it's out there and it's doin' things and you know, as a team leader I used to love it. It'd be like [laughs] drive it down there, do it, boom. Yay, I don't have to put the bomb suit on. Cool [laughs]. You know. Your operator, on the other hand, he's the guy that's gotta clean it, take care of it, fix it, put it together, you know, yeah, those types of things. But I … I love the robot. I think it's imperative. I think it's the greatest tool ever made for EOD operations.
Omar	Electro-mechanical device generally used to perform repetitive, dangerous or remote operations autonomously or under human guidance (or anywhere in between).	Basically, it's a mechanical extension to being something that … either because it's too repetitive, or too dangerous, that a human would normally do. It's kind of a labor-saving device. It's not directly connected to you like a Waldo would be, you know, where it replicates your gestures, but it could either be autonomous or-, or completely remote controlled, depending on the function. For bomb techs, of course, we want-, we like having some autonomous functions but we-, none of those autonomous functions are ones we want to start without us directing it. You know, talking to the [company redacted] guys and heavy

Name	Questionnaire	Interview
		engineers out in-, at many of our exercises, they wanna see how to build the product and they ask about, "Well, you know, what if you-, what if you did this or that?" And, you know, one of the things that we've said is that, "I don't care if the thing flips over; I don't want it trained to set itself upright without me telling it to." You know, it's great that it has the function that it'll automatically do that but ... if I don't tell it to do that, I don't want it to do it.
Wade	A remote-controlled machine that is controlled in some manner by a operator. It has some form of programmed or preset movements.	To me, it's some sort of remote-control machine that, it's gonna have programming, and so it has set functions, so both PackBot and TALON are perfect examples. There's three sets, so if you want it to go into ... I mean, if you want to recall the arm to its stowed position, I think the PackBot has a lot more as far as that because you can go into search modes and it automatically configures itself to what someone has decided is the best setup for that, or put it in travel mode. I don't think it'll brake itself yet.that's still, you have to learn how to do that with it. It's a type of machine, like I said, remote control that will have set programming so that you can say, do this, and it will do what you want it to do. It's either the radio controlled or tethered like our fiber optics.
Roy	A robot to me is an electronic or mechanical device.	A robot is a capable tool that allow you to do things from a remote position without exposing yourself to a hazard.
Simon	There are two definitions. One is it's a ... oh, what would you say? It's a mechanical invention designed to make our lives easier and safer. The other one is ... it's an extension of our own-, of our own personality. As they have to take on personality after you've used them for a while.	It's a mechanical invention designed to make our lives easier and safer. That's the main thing. And sometimes, it can be an extension of our own personality.
Isaiah	A mechanical tool that's used for many different purposes. It's being controlled by the most part; it can be controlled by a human.	Ah, so it's, to me, it would be, you know, a tool, which you know, any person, I mean, utilized to, you know, accomplish a specific task. I mean, could be used to entertain ... but, I mean, and it means, specifically in my experience, it's a tool that's utilized to keep, you know, the humans, you know, in a safe harbor. A very effective one, you know, which again, in the combat environment.

continued ...

Name	Questionnaire	Interview
Jed	A robot is a mechanical ... let's see ... I'd say it's a tool that allows an operator to do something from a distance.	A robot is a capable tool that allow you to do things from a remote position without exposing yourself to a hazard.
Reynaldo	Well, I guess, in my own words, it would just be, to me, it's ... a-, a tool used to accomplish, you know, a task ... for an EOD technician ... in a ha-, in a very hazardous or imminent threat situation.	A robot is just a ... a robot is a tool used, which really it, it's a tool for an EOD technician to use in a situation where there's an imminent threat, a high probability of a-, a detonation from an IED. It's a tool that he can use to remotely dispose of or RSP an IED. It's another tool ... . [laughs] to render safe.
Ben	A robot is a computerized machine or tool, I should say I guess, that when utilized with human interaction perform different maneuvers, tasks, accomplish goals, and everything dangerous when being controlled by a human.	It's just a machine, it's a tool, you know, that has a interface that has to have a human component mixed in with it in order to operate. I mean, in my instance, the use of word and how we use robots in EOD that was the nature of the thing now, I mean. There's other, you know, other definitions of robots. There's the sci-fi ones, the ones that they're developing in Japan right now, there's toy dog robots that, you know, for all intensive purposes [*sic*] I guess, really don't need any kind of human interaction, whereas as we used it ... as we used them and everything, it was a computerized machine that required human interface that allowed us to accomplish a specific goal with ourselves staying somewhat safe.
Leon	It's a tool.	A tool.
Rashad	A robot is a tool that humans operate.	I would say a mechanical, generally speaking, electronic and usually powered by a battery device that is normally used to accomplish some kind of mechanical task. But I'd also add to that, it could also just be used for ... auditory or visual acquiring. I don't think that would have been defined in my previous definition. You could just use a robot for a camera and the microphone. That'd be useful.
Quinn	A robot is a mechanical device with human interface that assists us in menial and/or dangerous tasks under direct supervision and control, you know, of the human operator.	It's a mechanical tool that is controlled by a human to do menial tasks and/or dangerous tasks to help alleviate the dangers of the human interface, i.e. put the mechanical piece of equipment downrange under direct supervision and control of the human.
Marcus	An electro-mechanical system that is either pre-programmed for a set of tasks, or controlled by people in order to help people.	A robot is an electro mechanical device that is utilized to help human beings to perform some kind of work or duty.

Name	Questionnaire	Interview
Mino	A robot is a device that's typically on battery that's wired, or usually wired, to a controlling unit that's controlled by a person or a human.	A robot's a tracked, battery-operated device that's operated by a human off of either a wireless or wired communication.
Connor	A robot is a remote platform designed to accomplish tasks from a remote location.	It's a system that allows you to accomplish tasks from a distance without putting boots near it. It's just a remote system to accomplish the same tasks, but safer.
David	A robot, overly extension of a, of my personal-, of my hands. The robot is a tool that we use, well, it's an amazing tool. It keeps-, it keeps us safe. Just … it's not very-, it can't work on its own, obviously, but it basically, if you could have the best robot in the world that can do anything and everything. But if you don't have an operator who is knowledgeable and who is trained on that piece of equipment, then that robot-, the robotics is pretty much obsolete.	Yeah, like I said before, it's just an extension of my hands. It's a tool we use and to keep people safe.
Axel	A robot's a machine that performs a task, but it's controlled by a human.	It's a machine that requires human input to accomplish a task.

Appendix B
US Navy EOD Ethos

I am a United States Navy EOD Technician, a warrior, professional Sailor and guardian of life.

I willfully accept the danger of my chosen profession and will accomplish all duties my great country asks of me.

I follow in the wake of those who have served before me with uncommon valor. I was born from the bombs and mines of the Blitzkrieg. I have cleared the world's sea lanes, and fought in the jungles, deserts, and mountains around the globe.

I will never disgrace the Navy EOD Warriors of the past and will uphold their honor and memory, both on and off the battlefield.

I am a quiet professional! I strive to excel in every art and artifice of war. I adapt to every situation and will overcome all obstacles. I will never fail those who depend on me.

I maintain my mind, body, and equipment in the highest state of readiness that is worthy of the most elite warrior.

I will defeat my enemies' spirit because my spirit is stronger. I will defeat my enemies' weapons because I know my enemies' weapons better.

I will complete every mission with honor, courage, and commitment. Though I may be alone and completely isolated, I will trust my teammates and my country. I will never give up and I will never surrender.

Where most strive and train to get it right, I will relentlessly train so I never get it wrong.

I am a United States Navy EOD Technician.

References

3d Explosive Ordnance Disposal (EOD) Battalion. (n.d.) On *Facebook* [government organization]. Retrieved October 29, 2011 from https://www.facebook.com/pages/3d-Explosive-Ordnance-Disposal-EOD-Battalion/183756438317677.

Abé, N. (2012, December 14). *Dreams in Infrared: The Woes of an American Drone Operator.* Spiegel Online International. Available at: http://www.spiegel.de/international/world/pain-continues-after-war-for-american-drone-pilot-a-872726.html.

Ackerman, E. (February 17, 2012). *DARPA wants to Give Soldiers robot Surrogates, Avatar Style.* Available at: http://spectrum.ieee.org/automaton/robotics/military-robots/darpa-wants-to-give-soldiers-robot-surrogates-avatar-style.

Afghanistan Annual Report on Protection of Civilians in Armed Conflict. (2012, March). Kabul, Afghanistan: United Nations Mission Assistance in Afghanistan (UNAMA) and Afghanistan Independent Human Rights Commission (AIHRC).

Air Land Sea Application Center. (2001, February). *EOD Multiservice Procedures for Explosive Ordnance Disposal in a Joint Environment.* Langley AFB, VA: ALSAC. Available at: General Dennis J. Reimer Training and Doctrine Digital Library: www.adtdl.army.mil.

Ambert, A., Adler, P.A., Adler, P., and Detzner, D.F. (1995, November) Understanding and evaluating qualitative research. *Journal of Marriage & the Family.* 57(4), 879–93. Available at: http://www.sociology.uwaterloo.ca/courses/soc712/ambert-adlers.pdf.

Arkin, R.C. (2005). Moving up the food chain: Motivation and emotion in behavior-based robots. In Fellous, J. and Arbib, M. (eds), *Who Needs Emotions: The Brain Meets the Robot.* New York: Oxford University Press.

Arkin, R.C. (2009). *Governing Lethal Behavior in Autonomous Robots.* Boca Raton, FL: CRC Press.

Arrow, H., McGrath, J.E., and Berdahl, J.L. (2000). *Small Groups as Complex Systems: Formation, Coordination, Development and Adaptation.* Thousand Oaks, CA: Sage.

Asimov, I. (1950). *I, Robot.* New York: Gnome Press.

Asimov, I. (1986). *Robot Dreams.* New York: Ace/Penguin Books.

Associated Press. (1989, August 31). Manny the robot helping to ensure safety of soldiers. *Daily News.* Available at: http://news.google.com/newspapers?id=8r QaAAAAIBAJ&sjid=HEgEAAAAIBAJ&pg=5321%2C7021877.

Associated Press. (14 September, 2007). Texas company builds emotional "robot boy.".

Atwood, M. (1996). *Alias Grace*. New York: Doubleday.

Auerbach, C.F. and Silverstein, L.B. (2003). *Qualitative Data: An Introduction to Coding and Analysis*. New York: New York University Press.

Axe, D. (2011, February 7). One in 50 troops in Afghanistan is a robot. *WIRED Magazine*. Available at: http://www.wired.com/dangerroom/2011/02/.

Axe, D. (2012, July 7). How to prevent drone pilot PTSD: Blame the 'bot. WIRED Magazine. Available at: http://www.wired.com/2012/06/drone-pilot-ptsd/.

Bailey, C. (2011, January). EOD Officer progression, diversity of EOD positions, and advantages of being dual tracked. *USAOC & S Newsletter*, 41(2), 7–8.

Barber, B. (1983). *The Logic and Limits of Trust*. New Brunswick, NJ: Rutgers University Press.

Bartneck, C., Reichenbach, J., and Carpenter, J. (2006).Well done robot! The importance of praise and presence in human-robot collaboration. *Proceedings of RO-MAN 06: The 15th IEEE International Symposium on Robot and Human Interactive Communication*, 86–90. Hatfield, UK. doi: 10.1109/ROMAN.2006.314414.

Bartneck, C., Reichenbach, J., and Carpenter, J. (2008). The carrot and the stick: The role of praise and punishment in human-robot collaboration. *Interaction Studies*, (9)2, 179–203. doi:10.1075/is.9.2.03bar.

Barylick, C. (2006, February 24). iRobot's PackBot on the front lines. *United Press International*. Available at: http://www.upi.com/Science_News/2006/02/24.

Bates, M.J. (2002, October). *Risk Factor Model Predicting the Relationship between Stress and Performance in Explosive Ordnance Disposal (EOD) training* (Doctoral dissertation). Available at: WorldCat Dissertations and Theses.

Bateson, G. (1972). *Steps to an Ecology of Mind: Collected Essays in Anthropology, Psychiatry, Evolution, and Epistemology*. Chicago, IL: University of Chicago Press.

Battarbee, K. and Mattelmaki, T. (2004). *Meaningful Product Relationships*. In D. McDonagh, P. Hekkert, J. Van Erp, and D. Gyi (eds), Design and Emotion). New York: Taylor & Francis, pp. 391–9.

Bazely, P. (2009). Mixed methods data analysis. In S. Andrews and E.J. Halcomb (eds), *Mixed Methods Research for Nursing and the Health Sciences*. Chichester: Wiley-Blackwell, pp. 84–114.

Belk, R.W. (1988). Possessions and the extended self. *Journal of Consumer Research*, 15(2), 139–68.

Bigelow, K. (Producer/Director). (2008). *The Hurt Locker* [Motion picture]. Universal City, CA: Summit Entertainment.

Billings, D.R., Schaefer, K.E., Kocsis, V., Barreram M., Cook, J., Chen, J.C.Y. (2012, March). *Human-Animal Trust as an Analog for Human-robot Trust: A Review of Current Evidence*. Aberdeen Proving Ground, MD: Army Research Laboratory.

Biocca, F., Harms, C., and Burgoon, J.K. (2003). Toward a more robust theory and measure of social presence: Review and suggested criteria, *Presence*, 12(5), pp. 456–80.

Blankenship, K. (2010, March 4). Face of defense: EOD experience benefits guard soldier. *U.S. Department of Defense News*. Available at: http://www.defense.gov/News/.

Blaustein, J. (Producer) and Wise, R. (Director). (1951). *The Day the Earth Stood still* [Motion picture]. United States: 20th Century Fox.

Bora, M. (2008, March 10). Jabil Circuit digs in on defense with robot: A St. Petersburg electronic design firm, in partnership with iRobot, ventures into the war zone with a machine that acts as a surrogate soldier in Iraq. *Tampa Bay Times*. Available at: http://www.sptimes.com/2008/03/10/news_pf/Business/Jabil_Circuit_digs_in.shtml.

Boston Dynamic. (2012, March). *PETMAN* [photograph]. Available at: http://www.bostondynamics.com/robot_petman.html.

Bowlby, J. (1973). *Attachment and Loss: Vol. 2., Separation: Anxiety and Anger.* New York: Basic Books.

Bowlby, J. (1980). *Attachment and Loss: Vol. 3., Loss: Sadness and Depression.* New York, NY: Basic Books.

Bowlby, J. (1982). *Attachment and Loss: Vol. 1. Attachment* (rev. ed.). New York: Basic Books.

Boyatzis, R. (1998). *Transforming Qualitative Information: Thematic Analysis and Code Development.* Thousand Oaks, CA: Sage.

Braun, V. and Clarke, V. (2006). Using thematic analysis in psychology. *Qualitative Research in Psychology*, 3, 77–101.

Breazeal, C., (2003, March 31). Toward sociable robots. *Robotics & Autonomous Systems*, 42, 167–75.

Breazeal, C. and Scassellati, B. (1999, October). How to build robots that make friends and influence people, *Intelligent Robots and Systems*, 1999. IROS '99. Proceedings from *IEEE/RSJ International Conference*, 2, 858–63. doi: 10.1109/IROS.1999.812787.

Bredo, E. (1994, Winter). Reconstructing educational psychology: Situated cognition and Deweyan pragmatism. *Educational Psychologist*, 29(1), 23–35.

Bronfenbrenner, U. (1979). *The Ecology of Human Development: Experiments by Nature and Design.* Cambridge, MA: Harvard University Press.

Brooks, R. (2002). *Robot: The Future of Flesh and Machines.* London: Penguin Books.

Brown, C. (2011, February 28). BigDog creator gets contracts for Cheetah and Atlas robots. *WIRED*. Available at: http://www.wired.co.uk/.

Brown, D. (2000). Defuse career doldrums: EOD wants you. *Navy Times*, 49(52), 18.

Bundy, E. and Sims, R. (2007, December). *Commonalities in an Uncommon Profession: Bomb Disposal.* Paper presented at the proceedings of the

Ascilite Singapore 2007, Singapore. Available at: http://www.ascilite.org.au/conferences/singapore07/procs/bundy.pdf.

Burgess, A. (1986). *A Clockwork Orange*. New York: W.W. Norton & Company.

Burke, J.L., Murphy, R., Rogers, E., Lumelsky, V.J., and Scholtz, J. (2004, May). NSF/DARPA study on human-robot interaction. Final report for the DARPA/ NSF interdisciplinary study on human-robot interaction. *Systems, Man, and Cybernetics, Part C: Applications and Reviews, IEEE Transactions*. 34(2), 103–12. doi: 10.1109/TSMCC.2004.826287.

Burton, B. (2014, September 2). *Japan's ISS Kirobo Robot is Lonely in Space*. C/Net. Available at: http://www.cnet.com/news/japans-iss-kirobo-robot-is-lonely-in-space/.

Campaign to Stop Killer Robots. (2013, July 13). *Report on Outreach on the UN Report on 'lethal autonomous robotics,'* [Report]. Wareham, M., Kastenson, K., Radejko, T., and Glidewell, S. (preparation). Washington, DC. Available at: http://stopkillerrobots.org/wp-content/uploads/2013/03/KRC_ReportHeynsUN_Jul2013.pdf.

Campion, M.A., Medsker, G.J., and Higgs, A.C. (1993). Relations between work group characteristics and effectiveness: Implications for designing effective work groups. *Personnel Psychology*, 46, 823–50.

Cantor, J. (2004). I'll never have a clown in my house: Why movie horror lives on. *Poetics Today*, 25(2), 283–304.

Čapek, K. (2004). *R.U.R. (Rossum's Universal Robots)*. (Novack-Jones, C., Trans.). New York: Penguin Classics. (Original work published 1920).

Carpenter, J. (2013). Just doesn't look right: Exploring the impact of humanoid robot integration into Explosive Ordnance Disposal teams. In R. Luppicini (ed.), *Handbook of Research on Technoself: Identity in a Technological Society*. Hershey, PA: Information Science Publishing, pp. 609–36. doi:10.4018/978-1-4666-2211-1.

Carpenter, J., Davis, J., Erwin-Stewart, N., Lee, T., Bransford, J., and Vye, N. (2008, October) *Invisible Machinery in Function, not Form: User Expectations of a Domestic Humanoid Robot*. Paper presented at the *proceedings of the 6th Conference on Design and Emotion*, Hong Kong, China.

Carpenter, J., Davis, J., Erwin-Stewart, N., Lee. T., Bransford, J., and Vye, N. (2009). Gender representation in humanoid robots for domestic use. *International Journal of Social Robotics (Special Issue*, 1875-4791), 1(3), 261–5. doi 10.1007/s12369-009-0016-4.

Carpenter, J., Eliot, M., and Schultheis, D. (2006, June). *The Uncanny Valley: Making Human-nonhuman Distinctions*. Paper presented at the *proceedings of the 5th International Conference on Cognitive Science*, Vancouver, BC. Available at: http://csjarchive.cogsci.rpi.edu/proceedings/2006/iccs/p81.pdf.

Carroll, C. (2012, April 2). Robots go from desert to jungle in new Navy lab. *Stars and Stripes*. Retrieved http://www.stripes.com/blogs/stripes-central/stripes-central-1.8040/robots-go-from-desert-to-jungle-in-new-navy-lab-1.173386.

Carroll, N. (2001). *On the Narrative Connection.* New perspectives on narrative perspective. Ed. Willie van Peer and Seymour Chatman. Albany, NY: SUNY Press.

Carter, J. (2002, 10 December). Nobel Peace Prize 2002 [speech]. Oslo, Norway. Available at: http://www.nobelprize.org/nobel_prizes/peace/laureates/2002/carter-lecture.html.

Casper, J. and Murphy, R.R. (2003). Human-robot interactions during the robot-assisted urban search and rescue response at the World Trade Center. *IEEE Transactions on Systems Man and Cybernetics Part B-Cybernetics*, 33(3), 367–85. doi:10.1109/TSMCB.2003.811794.

Chandler, J. and Schwarz, N. (2010). Use does not wear ragged the fabric of friendship: Thinking of objects as alive makes people less willing to replace them. *Journal of Consumer Psychology*, 20(2), 138–45. doi: 10.1016/j.jcps.2009.12.008.

Chapelle, W., Salinas, A, and McDonald, K. (April, 2011). *Psychological Health Screening of Remotely Piloted Aircraft (RPA) Operators and Supporting Units.* USAF School of Aerospace Medicine Department of Neuropsychiatry. Ohio, U.S.: Wright-Patterson Air Force Base.

Charmaz, K. (1991). *Good Days, Bad Days: The Self in Chronic Illness and Time.* New Brunswick, NJ: Rutgers University Press.

Charmaz, K. (2000). *Grounded Theory: Objectivist and Constructivist Methods.* In N.K. Denzin and Y.S. Lincoln (eds), *Handbook of Qualitative Research* (2nd Edition, pp. 509–35). Thousand Oaks, CA: Sage.

Chatfield, J.A. (1995, June). *Force Feedback for Anthropomorphic Teleoperated Mechanism* (Master's thesis). United States Navy, Naval Postgraduate School: Monterey, CA.

Chen, J.C.Y. (1996). *Early Chinese Work in Natural Science: A Re-examination of the Physics of Motion, Acoustic, Astronomy, and Scientific Thought.* Hong Kong: Hong Kong University Press.

Choate, H. (2011). *EOD Marines Honor Fallen Comrades with Memorial Wall.* Available at: Defense Video & Imagery Distribution System website: http://www.dvidshub.net/news/66453.

Cohen, L., Manion, L., and Morrison, K. (2007). *Research Methods in Education* (6th ed.). New York: Routledge.

Cooper, J. (2011, August 10). Langley EOD: Serving at home and overseas. *U.S. Air Force News.* Available at: http://www.acc.af.mil/news/story.

Cullins, A. (2011, December 9). Advocates fight for classification of "four-footed soldiers." *Medill National Security Zone.* Available at: http://nationalsecurityzone.org/site/advocates-fight-for-reclassification-of-four-footed-soldiers/.

Darken, R., Kempster, K., and Peterson, B. (2001, October). *Effects of Streaming Video Quality of Service on Spatial Comprehension in a Reconnaissance Task.* Paper presented at the proceedings of the Meeting of I/ITSEC, St. Louis, MO.

DARPA. (n.d.). *ARM: Autonomous Robot Manipulation.* Available at: http://thearmrobot.com/.

DeCuir-Gunby, J., Marshall, P., and McCulloch, A. (2011). Developing and using a codebook for the analysis of interview data: An example from a professional development research project. *Field Methods*, 23(2), 136–55.

DeLillo, D. (1999). *Underworld.* London: Pan Macmillan.

DeYoung, D. (1983). *Mr. Roboto* [Styx]. On *Kilroy was here* [music]. Santa Monica, CA: A&M.

Dennen, R. (2011, September 16). Explosives school puts on display. *Fredericksburg.com.* Available at: http://fredericksburg.com/News/FLS/2011/092011/09162011.

Denzin, N.K. (1989). *The Research Act: A Theoretical Introduction to Sociological Methods* (3rd ed.). Englewood Cliffs, NJ: Prentice Hall.

Denzin, N.K. (2006). *Sociological Methods: A Sourcebook* (5th ed.). Piscataway, NJ: Transaction Publishers.

Department of Defense. (2006). *Urban Operations plus Explosive Ordnance Disposal Multiservice Procedures for EOD in a Joint Environment (FM-306).* Washington, DC: Pentagon Publishing.

Dietz, S. (2014, July 12). *Meeting LS3: Marines Experiment with Military Robotics.* Defense Video and Imagery Distribution System (DVIDS). Available at: http://www.dvidshub.net/news/135952/meeting-ls3-marines-experiment-with-military-robotics.

Dirks, K.T. and Ferrin, D.L. (2001). The role of trust in organizational settings. *Organization Science*, 12(4), 450–67.

DiSalvo, C., Gemperle, F., Forlizzi, J., and Kiesler, S. (2002, June). *All Robots are not Created Equal: The Design and Perception of Humanoid Robot Heads.* Paper presented at the proceedings of DIS2002, London, UK. doi:10.1145/778712.778756.

Domegan, C. and Fleming, D. (2007). *Marketing Research in Ireland: Theory and Practice.* New York: Gill & MacMillan.

Doyle, S. (2007). Member checking with older women: A framework for negotiating meaning. *Health Care for Women International*, 8(10), 888–908.

Dreazen, Y.J. (2011, March 3). IED casualties up despite increased vigilance: Military's outgoing head of IED-combating task force says insurgents will continue to use the cheap, deadly weapons. *The National Journal.* Available at: http://www.nationaljournal.com/.

Duffelblog. (2013, October 26). *EOD Soldier Files Discrimination suit after Army Denies Marriage to his Robot.* Available at: http://www.duffelblog.com/2013/10/eod-soldier-files-discrimination-suit-army-denies-marriage-loved-robot/.

Duffy, B.R. (2000). *The Social Robot.* (Doctoral dissertation). Available at: WorldCat Dissertations and Theses.

Duffy, T.M. and Cunningham, D.J. (1996). Constructivism: Implications for the design and delivery of instruction. In D.H. Jonassen (ed.). *Handbook for Research in Educational Communications*. New York: MacMillan, pp. 170–99.

Dunn, M.W. (1995). *A Theory of Animate Perception*. (Doctoral dissertation). Available at: WorldCat Dissertations and Theses.

Dyess, B.G., Winstead, M. and Golson, E. (2011, March 31). *The Role of Unmanned Systems in the Army*. [PDF document]. Army Capabilities Information Center.

Edwards, L. (2010). *PETMAN robot to Closely Simulate Soldiers*. Available at: http://www.physorg.com/news191563032.html.

Eisenberg, J. (2007). Group cohesiveness. R.F. Baumeister and K.D. Vohs (eds), *Encyclopaedia of Social Psychology*. Thousand Oaks, CA: Sage, pp. 386–8.

EOD Ethos. (2013). (N.A.). United States Navy website. Available at: http://www.public.navy.mil/bupers-npc/officer/communitymanagers/Unrestricted/Documents/EOD%20Ethos.pdf.

EOD Memorial Foundation. (2009, December 1). *EOD Memorial Placement guidelines* [Memo]. Niceville, FL: Author.

EOD Memorial Foundation. (2011). *Scholarship*. Available at: http://www.eodmemorial.org/about/.

EOD Memorial Foundation. (n.d.). *About: History of the Memorial*. Available at: http://www.eodmemorial.org/about/.

Epley, N., Akalis, S., Waytz, A., and Cacioppo, J.T. (2008). Creating social connection through inferential reproduction: Loneliness and perceived agency in gadgets, gods, and greyhounds. *Psychological Science*, 19(2), 114–20. doi: 10.1111/j.1467-9280.2008.02056.x.

Epley, N., Waytz, A., Akalis, S., and Cacioppo, J.T. (2008). When we need a human: Motivational determinants of anthropomorphism. *Social Cognition*, 26(2), 143–55.

Ernest, P. (1998). *Social Constructivism as a Philosophy of Mathematics: Radical Constructivism*. Albany, NY: State University of New York Press.

Explosive Ordnance Disposal Group 1. (2013, January 18). In *Facebook* [U.S. Navy group]. Available at: https://www.facebook.com/EODGROUP1.

Everett, H.R., Pacis, E.B., Kogut, G., Farrington, N., and Khurana, S. (2004, October 26). *Towards a Warfighter's Associate: Eliminating the Operator Control Unit*. Proceedings of SPIE 5609: Mobile Robots XVII. Ft. Belvoir: Defense Technical Information Center. doi: 10.1117/12.571458.

Favre, D. (2010). *Living Property: A New Status for Animals within the Legal System*. Marquette Legal Review, 93, 1021–71.

Feigenbaum, E.A. and McCorduck, P. (1982). The fifth generation: Artifical intelligence and and Japan's computer challenge to the world. Reading, MA: Addison-Wesley.

Fiddian, P. (2012, July). *Smart Suit Could Boost Troop Performance*. Armed Forces International News. Available at: http://www.armedforces-int.com/news/smart-smart-could-boost-troop-performance.html.

Fincannon, T., Barnes, L., Murphy, R.R., and Riddle, D.L. (2004, September -October*). Talking and Gesturing to a Robot: Emergent Social Interaction in Rescue Robots*. Paper presented at the proceedings of the 2004 IEEE/ RSJ International Conference on Intelligent Robots and Systems (IROS), Piscataway, NJ.

Finkelstein, R. and Albus, J. (2003, May 13; revised 2004, November). *Technology Assessment of Autonomous Intelligent Bipedal and Other Legged Robots*. Arlington, VA: Defense Advanced Research Project Agency.

Fisher, A. (1988, September). Sweaty Manny. *Popular Science*, 233(3), 10. Available at: http://www.popsci.com/archive.

Flaherty, A. (2010, April 8). IEDs in Afghanistan double in past year. *Associated Press*. Available at: http://www.armytimes.com/news/.

Fong, T., Nourbakhsh, I. and Dautenhahn, K. (2003). A questionnaire of socially interactive robots. *Robotics and Autonomous Systems*, 42(3-4), 143–66.

Forsyth, D.R. (2010). Cohesion and development. In *Group Dynamics* (5th ed.). Wadsworth: Cengage Learning, pp. 116–42.

Fowler, F.J. (1988). *Survey Research Methods*. Beverly Hills, CA: Sage Publications.

Frankfort, H., Frankfort, H.A., Jacobsen, T. and W.A. Irwin. (1977). *The Intellectual Adventure of Ancient Man: An Essay on Speculative Thought in the Ancient Near East*. Chicago: University of Chicago Press.

Freud, S. (1990). The Uncanny. *The Penguin Freud Library Volume 14: Art and Literature*. (Trans. and ed. J. Strachey) London: Penguin. (Original work published 1919).

Fussell, S.R., Kiesler, S., Setlock, L.D., and Yew, V. (2008, March). *How People Anthropomorphize Robots. Proceedings of Conference of Human-Robot Interaction HRI '08*, Amsterdam, Netherlands. doi: 10.1145/1349822.1349842.

Garreau, J. (2007, May 6). 'Bots on the ground: In the field of battle (or even above it), robots are a soldier's best friend. *The Washington Post*. Available at: http://www.washingtonpost.com.

Geertz, C. (1973). The cerebral savage: On the work of Claude Lévi-Strauss. In *The Interpretation of Cultures: Selected Essays*. New York: Basic Books. Available at: Ann Arbor: University Microfilms International, p. 362.

Geertz, C. (1975). Thick description: Toward an interpretive Theory of Culture. *In The Interpretations of Cultures: Selected Essays*. London: Hutchinson, pp. 3–30.

Gergen, K.J. (1985). The social constructionist movement in modern psychology. *American Psychologist*, 40(3), 266–75.

Gibson, F.M. (2009, August 14). *US Navy EOD Experts Train Philippine Navy SEAL team*. Joint Special Operations Task Force: Philippines. Available at: http://jsotf-p.blogspot.com/2009/08/.

Gibson, W. (2005). *Pattern Recognition*. New York: Penguin.

Gilbert, G.R. and Beebe, M.K. (2010). *United States Department of Defense Research in Robotic Unmanned Systems for Combat Casualty Care*. U.S. Army Medical Research and Materiel Command Telemedicine and

Advanced Technology Research Center. Ft. Detrick, MD: Defense Technical Information Center.

Glaser, B.G. and Strauss, A.L. (1967). *Discovery of Grounded Theory: Strategies for Qualitative Research.* Chicago, IL: Aldine.

Godwin, W. (1876). *Lives of the Neuromancers: Or, an Account of the Most Eminent Persons in Successive Ages who have Claimed for Themselves or to whom has been Imputed by Others the Exercise of Magical Powers.* London: Chatto & Windus.

Goldeberg, L., Hashimato, R., Schneider, H., McNall, B., (Producers) and Badham, J. (Director). (1983). *War Games* [Motion picture]. USA; MGM/ UA Entertainment.

Goldsmith, A.L. (1981). *The Golem Remembered 1909-1980: Variations of a Jewish Legend.* Detroit, MI: Wayne State University Press.

Gonzales, R.T. (2013, October 28). *Psychologists Propose Horrifying Solution to PTSD in Drone Operators. iO9.* Available at: http://io9.com/psychologists-propose-horrifying-solution-to-ptsd-in-dr-1453349900/all.

Goodwin, D., Pope, C., Mort, M., and Smith, A. (2003). Ethics and ethnography: An experiential account. *Qualitative Health Research,* 13, 567–77.

Gray, C.H. (1995). *The Cyborg Handbook.* New York: Routledge.

Gredler, M.E. (1997). *Learning and Instruction: Theory into Practice (3rd ed).* Upper Saddle River, NJ: Prentice Hall.

Groom, V., Takayama, L., Ochi, P., and Nass, C. (2009, March). *I am my Robot: The Impact of Robot-building and Robot Form on Operators.* Paper presented at the proceedings of the 4th ACM/IEEE International Conference on Human Robot Interaction, La Jolla, CA. doi:10.1145/1514095.1514104.

Guizzo, E. (2010, October 18). *DARPA Seeking to Revolutionize Robotic Manipulation.* IEEE Spectrum InsideTechnology. Available at: http://spectrum. ieee.org/automaton/robotics/.

Hall, K. (2011, April 1). Army bomb disposal units train at Campbell. *Army News.* Available at: http://www.armytimes.com/news/2011/04/ap-army-bomb-disposal-units-compete-at-campbell-040111/.

Hamilton, H. (2004). *The Speckled People.* London: Harper Perennial.

Hancock, P.A., Billings, D.R., and Schaefer, K.E. (2011). Can you trust your robot? *Ergonomics in Design,* 19, 24–29.

Hapgood, F. (2008, June). When robots live among us. *Discover Magazine.* Available at: http://discovermagazine.com/2008/jun/27-when-robots-live-among-us.

Hartman, J.J. CAPT Stewart and Lucas [Photograph]. (2012). *Navy Sets Sail with Robotics Lab.* By Martin LaMonica, CNET. Available at: http://news.cnet. com/8301-11386_3-57408713-76/navy-sets-sail-with-robotics-lab/.

Hay, K. personal communication, October 2, 2013.

Heider, F. (1958). *The Psychology of Interpersonal Relations.* Hillsdale, NJ: Lawrence Erlbaum.

Heider, F. and Simmel, M. (1944). An experimental study of apparent behavior. *The American Journal of Psychology*, 243–59.

Henderson, Z. (1961). *Pilgrimage: The Book of the People*. New York: Avon Books.

Hinds, P.J., Roberts, T.L., and Jones, H. (2004). Whose job is it anyway?: A study of Human-Robot Interaction in a collaborative task. *Human-Computer Interaction*, 19, 151–81.

Hogan, J. and Hogan, R. (1989). Noncognitive predictors of performance during explosive ordnance disposal training. *Military Psychology*, 1, 117–33.

Human Rights Watch. (2012). *Ban 'Killer Robots' before it's too late: Fully Autonomous Weapons would Increase Danger to Civilians*. [Press release]. Available at: http://www.hrw.org/news/2012/11/19/ban-killer-robots-it-s-too-late.

Hurd, G.A. (Producer) and Cameron, J. (Director). (1984). *The Terminator (Special Edition)* [Motion picture]. United States: MGM.

Hurtado, R. (2014, 27 March). Marine finds strength in furry companion. *Military. com News*. Available at:: http://www.military.com/daily-news/2014/03/27/marine-finds-strength-in-furry-companion.html.

Hyginus, G.J. (1960). The myths of Hyginus. (M. Grant, Trans.). *Humanistic Studies, no. 34*. Lawrence, KS: University of Kansas Press. (Original work published c. 900).

iCasualties.org. (2011). *IED Fatalities*. Available at: http://icasualties.org/OEF/index.aspx.

Idaho National Laboratory. (1988). Manny [photograph]. Available at: https://inlportal.inl.gov/portal/server.pt/community/historical_perspectives/537.

Idel, M. (1990). *Golem: Jewish Magical and Mystical Traditions on the Artificial Anthropoid*. Albany, NY: State University of New York Press.

iRobot. (2013). 710 Warrior [photograph]. Available at: http://www.irobot.com/en/us/robots/defense/warrior/Details.aspx.

Janowitz, M. (1972). Characteristics of the military environment. In S.E. Ambrose and J.A. Barber (eds), *The Military and American Society: Essays and Reading*. New York: Free Press, pp. 166–72.

Jean, G.V. (2011, July). New robots planned for bomb disposal teams. *National Defense Magazine*. Available at: http://www.nationaldefensemagazine.org/archive/2011/July/Pages/NewRobotsPlannedforBombDisposalTeams.aspx.

JIEDDO. (2012a). *Counter-Improvised Explosive Device Plan: 2012-2016*. Available at: https://www.jieddo.mil/content/docs/20120116_JIEDDOC-IEDStrategicPlan_MEDprint.pdf.

JIEDDO. (2012b, August). *Global IED Monthly Summary Report*. Available at: http://info.publicintelligence.net/JIEDDO-MonthlyIEDs-AUG-2012.pdf.

Jiminez, J.S. (2011). RSP: The newsletter of the National EOD Association. *Commander's Message*, 1, 1.

Johns, J.H., Bickel, M.D., Blades, A.C., Creel, J.B., Gatling, W.S., Hinkle, J.M., and Stocks, S.E. (1984). *Cohesion in the U.S. Military*. Fort Lesley J. McNair,

Washington, DC: National Defense University Press. Available at: http://handle.dtic.mil/100.2/ADA140828.

Jones, H. and Hinds, P. (2002, November). *Extreme Work Teams: Using SWAT Teams as a Model for Coordinating Distributed Robots*. Paper presented at the proceedings of 2002 ACM Conference on Computer Supported Cooperative Work, New Orleans, LA. doi:10.1145/587078.587130.

Kambayashi, S. (June 18, 2015). Japanese firm prepares to sell line of childlike robots with 'emotional' responses. *The Orange County Register*. Available at: http://www.ocregister.com/articles/son-667376-robot-pepper.html.

Kane, G. (2014, May 8). Sargeant Stubby: America's original war dog fought bravely on the Western Front, then helped the nation forget the Great War's terrible human toll. *Slate Magazine*. Available at: http://www.slate.com/articles/news_and_politics/history/2014/05/dogs_of_war_sergeant_stubby_the_u_s_army_s_original_and_still_most_highly.single.html.

Komarow, S. (2005, October 25). Robots nail down the nuts and bolts of bomb disposal. *USAToday*, 10A. Available at: http://usatoday30.usatoday.com/tech/news/techinnovations/2005-10-24-robotwar_x.htm.

Kelly, M. and Johnson, R. (2012, August 15). This is what disarming bombs in the military is really like. *Business Insider*. Available at: http://www.businessinsider.com/an-eod-technician-explains-what-life-is-really-like-in-the-field-2012-8?op=1.

Kelsey, A. (2011, July 1). First Airmen graduate from new EOD screening course. *U.S. Air Force News*. Available at: http://www.af.mil/news/.

Kennedy, K., Spielberg, S., and Curtis, B. (Producers) and Spielberg, S. (Director). (2001). *AI* [Motion picture]. United States: Amblin Entertainment.

Kidd, C. and Breazeal, C. (2005, April). Human-robot interaction experiments: Lessons learned. *Proceedings of AISB 2005*. United Kingdom: University of Hertfordshire.

King, N. and Horrocks, C. (2010). *Interviews in Qualitative Research*. Thousand Oaks, CA: Sage.

Kirke, C. (January 01, 2009). Group cohesion, culture, and practice. *Armed Forces & Society*, 35(4), 745–53. doi: 10.1177/0095327X09332144.

Kokakaya, S. (2010, January). An educational dilemma: Are educational experiments working? *Education Research & Review*. Available at: http://www.academicjournals.org/err/PDF/Pdf%202011/Jan/Kocakaya.pdf.

Kolb, M. (2012). *Soldier and Robot Interaction in Combat Environments*. (Doctoral dissertation). Available from ProQuest Dissertations and Theses databse. (UMI No. 3524365). Available at: http://gradworks.umi.com/35/24/3524365.html.

Kukla, A. (2000). *Social Constructivism and the Philosophy of Science*. New York: Routledge.

Kurtz, G. (Producer), and Lucas, G. (Director). (1977). *Star Wars* [Motion picture]. United States: LucasFilm.

Kurzweil, R. (1999). *The Age of Spiritual Machines: When Computers Exceed Human Intelligence*. New York: Viking.

Lakoff, G. (1992). The contemporary theory of metaphor. In Ortony, A. (ed.), *Metaphor and Thought*, 2 ed. New York: Cambridge University Press.

Lamance, R. (2010). *Grandmother Graduates from Explosive Ordnance School.* U.S. Department of Defense. San Antonio, TX: Defense Media Acitivity. Available at: http://www.defense.gov/News/NewsArticle.aspx?ID=60811.

Larry, D.A. (2008). Bomb squads and EOD personnel: Interoperability for Homeland Defense. *Army Logistician*, 40(3). Available at: http://www.almc. army.mil/alog/issues/MayJun08/ bomb_eodpersonnel.html.

Lazarus, R.S. (1966). *Psychological Stress and the Coping Process*. New York: McGraw-Hill.

Lazarus, R.S. (1993). From psychological stress to the emotions: A history of changing outlooks. *Annual Review of Psychology*, 44, 1–21.

Lazarus, R.S. (2006). *Stress and Emotions: A New Synthesis*. New York: Springer Publishing Company.

Lazarus, R.S. and Folkman, S. (1984). *Stress, Appraisal, and Coping*. New York: Springer.-.

Lee, S.I., Kiesler, S., Lau, I.Y., and Chiu, C.Y. (2005). Human mental models of humanoid robots. *Proceedings*, 3, 2767–72.

LeRoy, M. (Producer) and Fleming, V. (Director). (1939). *The Wizard of Oz*. [motion picture]. USA: MGM.

Lewis, M., Wang, J., and Hughes, S. (2007). USARSim: Simulation for the study of Human-Robot Interaction. *Journal of Cognitive Engineering and Decision Making*, 1(1), 98–120.

Levy, D. (2007). *Love + Sex with Robots*. New York: HarperCollins.

Lin, P., Bekey, G. and Abney, K. (2008, December 20). *Autonomous Military Robotics: Risk, Ethics and Design*. U.S. Department of Navy, Office of Naval Research. San Luis Obispo, CA: California Polytechnic State University, San Luis Obispo.

Lincoln, Y.S. and Guba, E.G. (1985). *Naturalistic Inquiry*. Beverly Hills, CA: Sage.

Linebaugh, H. (2013, December 29). I worked on a US drone program: The public should know what really goes on. *The Guardian*. Available at: http://www. theguardian.com/commentisfree/2013/dec/29/drones-us-military.

Livingstone, D.W. (2001). *Adults' Informal Learning: Definitions, Finds, Gaps, and Future Research: New Approaches for Lifelong Learning* (NALL), Working paper # 21-2001, Toronto, Ontario, Canada.

Lowers, G. (2013, February 21).Quiet professional receives Purple Heart medal. *U.S. Army News*. Available at: http://www.army.mil/article/96947/.

Lumière, A. and Lumière, L. (Producers), Lumière, A. and Lumière, L. (Directors). 1896. *L'arrivée d'un train en gare de La Ciotat*. [Motion picture]. France: Société Lumière.

MacCoun, R.J. and Hix, W.M. (2010). Unit cohesion and military performance. In National Defense Institute (collective authorship), *Sexual Orientation and U.S. Military Policy: An Update of RAND's 1993 Study*. Santa Monica, CA.: RAND, pp. 137–65.

Magnuson, S. (2009, March). Humanoid soldiers: Reverse engineering the brain may accelerate robotics research. *National Defense*, 32–4.

Mark, G. (1997, November). *Merging Multiple Perspectives in Groupware Use: Intra- and Intergroup Conventions*. Paper presented at the proceedings of the International ACM SIGGROUP Conference on Supporting Group Work: The Integration Challenge, Phoenix, AZ. doi:10.1145/266838.266846.

Mark, L., Davis, J., Dow, T., and Godfrey, W. (Producers), and Proya, A. (Director). (2004). *I, Robot* [Motion picture]. USA: 20th Century Fox.

McCormick, T. (2014, January 24). *Foreign Policy*. Available at: http://foreignpolicy.com/2014/01/24/lethal-autonomy-a-short-history.

McCracken, G. (1988). *The long interview*. Newbury Park, CA: Sage.

McEntee, P. (1989, September 1). 'Manny' tests military garb: Robot's mission ensures protection for soldiers. *Schenectady Gazette*. Available at: http://news.google.com/newspapers?id=5GlGAAAAIBAJ&sjid=1ugMAAAAIBAJ&pg=983%2C19011.

McGrath, J.E. (1976). Stress and behavior in organizations. In M.D. Dunnette (ed.), *Handbook of Industrial and Organizational Psychology*. Chicago, IL: Rand McNally, pp. 1351–95.

McKinney, D. (2012). NRL designs robots for shipboard firefighting. Available at: http://www.nrl.navy.mil/media/news-releases/2012/nrl-designs-robot-for-shipboard-firefighting.

McMahon, M. (1997, December). Social constructivism and the World Wide Web: A paradigm for learning. *Proceedings of ASCILITE conference*, Perth, Australia.

Merriam, S.B. (1988). *Case Study Research in Education: A Qualitative Approach*. San Francisco, CA: Jossey-Bass Publishers.

Miles, M. and Huberman, M. (1994). *Qualitative Data Analysis: A Sourcebook of New Methods* (2nd ed.). Newbury Park, CA: Sage.

Millsaps, B.B. (2015, March 30). *Lithuanian Programmer 3D Prints Robotic Tank Replica from Sci-fi Movie 'Ghost in the Shell.'* 3dprint.com. Available at: http://3dprint.com/54671/ghost-in-the-shell-robot-tank/.

Mizuo, Y., Matsumoto, K., Iyadomi, K. and Ishikawa, M. (Producers) and Oshii, M. (Director). (1995). *Ghost in the Shell* [Motion picture]. Japan: Production I.G.

Mora, E. (2010, December 8). IEDs being used against stable governments worldwide, says top U.S. Military official. *CNSC News*. Available at: http://www.cnsnews.com/.

Mori, M. (2012). The Uncanny Valley. (K.F. MacDorman and N. Kageki, Trans.) *Energy*, 7(4), 33–5. (Original work published 1970).

Morris, J. (Producer) and Stanton, A. (Director). (2008). *Wall-E [Theatrical Release]* [Motion picture]. United States: Pixar Animation Studios.

MOS 89D: Explosive Ordnance Disposal Specialist. (n.d.) *U.S. Army Info*. Available at: http://www.us-army-info.com/pages/mos/ordnance/55d.html.

Muir, B.M. and Moray, N. (1996). Trust in automation. Part II. Experimental studies of trust and human intervention in a process control simulation. *Ergonomics*, 39(3), 429–60.

Murphy, R.R. (2000). *Introduction to AI Robotics*. Cambridge, MA : MIT Press.

Murphy, R.R. (2004). Human-robot interaction in rescue robotics. *IEEE Transactions on Systems, Man and Cybernetics: Part C–Applications and Reviews*, 34(2), 138–53. doi:10.1109/TSMCC.2004.826267.

Murphy, R.R. and Woods, D.D. (2009). Beyond Asimov: The three laws of responsible robotics. *Intelligent Systems, IEEE*, 24(4), 14–20.

Nass, C. and Moon, Y. (2000). Machines and mindlessness: Social responses to computers. *Journal of Social Issues*, 56(1), 81–103.

National EOD Association. (2012, July). *RSP: The Newsletter of the National EOD Association*. Available at: http://www.nateoda.org/PDF_Files/RSPJUL2012. pdf.

National Public Radio (Producer). (2012, November 30). *This American Life: (480) Animal Sacrifice*. [Interview transcript]. Available at: http://www. thisamericanlife.org/radio-archives/episode/480/animal-sacrifice.

National Research Council. (2002). *Making the Nation Safer: The Role of Science and Technology in Countering Terrorism*. Washington, DC: National Academy Press.

Naval Explosive Ordnance School. (2001). Available at: Eglin Air Force Base website: http://www.united-publishers.com/EglinGuide/units.html.

Naylor, J.C. and Dickenson, T.L. (1969). Task structure, work structure, and team performance. *Journal of Applied Psychology*, 53, 167–77.

Nerlich, B., Clarke, D.D. and Dingwall, R. (2008, March). Fictions, fantasies and fears: The literary foundations of the cloning debate. *Journal of Literary Semantics*, 30(1), 37–52. doi:10.1515/jlse.30.1.37.

Nieva, V.F., Fleishman, E.A., and Rieck, A. (1978). *Team Dimensions: Their Identity, their Measurement, and their Relationships (Technical Report, Contract No. DAHC19-78-C-0001)*. Washington, DC: Advanced Research Resources Organizations.

Nitsche, E. [Illustrator]. (July 28, 1935). *When Wars are Fought with Robot Soldiers*. San Antonio Light, p. 82. Retrieved from http://newspaperarchive. com/us/texas/san-antonio/san-antonio-light/1935/07-28/page-82.

Norman, D.A. (1988). *The Design of Everyday Things*. New York: Doubleday.

Norman, D.A. (2004). Emotional design: Why we love (or hate) everyday things. New York: Basic Books.

Norman, D.A. (2005, March 1). Robots in the home: What might they do? *Interactions*, 12(2), 65.

Osborn, K. (2010, October 28). Army building smarter robots. *Army News Service*. Available at: http://www.army.mil/article/47344/.

Ovid. (2009, November). *Metamorphoses*. (C. Martin, Trans.). New York: W.W. Norton & Company. (Original work published A.D. 8).

Owolabi, O. (2010, January 4). Airmen train Kyrgyz officials on EOD mission. *Air Force Print News Today*. Available at: http://www.af.mil/news/story. asp?id=123184049.

Palinscar, A.S. (1998). Social constructivist perspectives on learning and teaching. *Annual Review of Psychology*, 49(1), 345–75.

Palinscar, A.S. and Herrenkohl, L.R. (2002, Winter). Designing collaborative learning contexts. *Theory into Practice*, 41(1). doi:10.1207/ s15430421tip4101_5.

Pantozzi, J. (2012, May 10). *This Mechanical Engineer had a Bomb Disposal Robot as Ring-bearer at her Wedding*. The Mary Sue. Available at: http://www. themarysue.com/robot-ring-bearer/.

Patton, M.Q. (1990). *Qualitative Evaluation and Research Methods* (2nd ed.). Thousand Oaks, CA: Sage.

Power, M. (2013, October 23). *Confessions of a Drone Warrior*. GQ. Available at: http://www.gq.com/news-politics/big-issues/201311/drone-RPA-pilot-assassination.

Prawat, R.S. and Floden, R.E. (1994). Philosophical perspectives on constructivist views of learning. *Educational Psychologist*, 29(1), 37–48.

QinetiQ North America. [Photograph of TALON IV]. (2011). Available at: https:// www.facebook.com/qinetiqnarobots/photos_stream.

Rafaeli, A. and Vilnai-Yavetz, I. (2004). Emotion as a connection of physical artifacts and organizations. *Organization Science*, (15)6, 671–86. doi:10.1287/ orsc.1040.0083.

Reeves, B. and Nass, C. (1996). *The Media Equation: How People Treat Computers, Televisions, and New Media like Real People and Places*. Stanford, CA: Cambridge University Press.

Rempel, J.K., Holmes, J.G., and Zanna, M.P. (1985). Trust in close relationships. *Journal of Personality and Social Psychology*, 49(1), 95.

Riemer, C.F. (2008, May 22). *The Organizational Implications of the U.S. Army's Increasing Demand for Explosive Ordnance Disposal Capabilities*. [Monograph]. School of Advanced Military Studies, United States Army Command and General Staff College. Ft. Belvoir: Defense Technical Information Center. Available at: http://www.dtic.mil/dtic/tr/fulltext/u2/ a485658.pdf.

Rizzo, J. (2012, January 6). When a dog isn't a dog. *CNN*. Available at: http:// security.blogs.cnn.com/2012/01/06/when-a-dog-isnt-a-dog/.

Robillard, T.K. (2011, March 15). Picatinny advances EOD training with video game technology. *Army News Service*. Available at: http://www.army.mil/ article/53259/.

Robotic Systems Joint Project Addendum: Unmanned Ground Systems Roadmap. (2012, July). N.A. Available at: http://www.rsjpo.army.mil/images/UGS_ Roadmap_Addendum_Jul12.pdf.

Roderick, I. (2010). Considering the fetish value of EOD robots: How robots save lives and sell war. *International Journal of Cultural Studies*, 13(3), 235–53.

Rogoff, B. (1990). Apprenticeship in thinking: Cognitive development in social context. New York: Oxford University Press.

Rose, B. (2011, December 28). *The Sad Story of a Real Life R2-D2 who Saved Countless Human Lives and Died.* GIZMODO. Available at: http://gizmodo.com/5870529/.

Rosenman, M.S. and Gero, J.S. (1998). Purpose and function in design: From the socio-cultural to the techno-physical. *Design Studies*, 19(2), pp. 161–86.

Rosenthal-von der Pütten, A.M., Krämer, N.C., Hoffmann, L., Sobieraj, S., and Eimler, S.C. (2012). An experimental study on emotional reactions towards a robot. *International Journal of Social Robotics*, 1–18.

Ross, L.D. (1977). The intuitive psychologist and his shortcomings: Distortions in the attribution process. In L. Berkowitz (Ed.), *Advances in Experimental Social Psychology* (Vol. 10, pp. 173–220). New York: Academic Press.

Rousseau, D.M. and Cooke, R.A. (1988, August). *Cultures of High Reliability: Behavioral Norms Aboard a U.S. Aircraft Carrier.* Paper presented at the meeting of the Academy of Management, Anaheim, CA.

Ryle, G. (1968). The thinking of thoughts: What is Le Penseur doing? *University Lectures, 18.* University of Saskatchewan. Available at: http://lucy.ukc.ac.uk/CSACSIA/Vol14/Papers/ryle_1.html.

St. Nicholas Magazine. (1875, May). Unknown author. 2(7), pp. 448–49.

Sanchez, S. (2005, January 28). Command assesses robot to help save soldiers' lives. U.S. Department of Defense. Retrieved http://www.defense.gov/transformation/articles/2005-01/ta012505a.html.

Scandura, T.A. and Williams, E.A. (2000). Research methodology in management: Current practices, trends, and implications for future research. *Academy of Management Journal*, 43, 1248–64.

Scarborough, R. (2012, June 3).Delta Force: Army's "quiet professionals" operate in shadows – not in spotlight. *The Washington Times.* Available at: http://www.washingtontimes.com/news/2012/jun/3/.

Scholtz, J.C. (2002, March). Creating synergistic cyber forces. Alan C. Schultz and Lynne E. Parker (eds), Multi-Robot Systems: From Swarms to Intelligent Automata. *Proceedings of 2002 NRL Workshop on Multi-Robot Systems*, Washington, DC: Kluwer Academic Publishers, pp. 177–84.

Scholtz, J. (2003, January). Theory and evaluation of human-robot interactions. *Proceedings of 2003 International Workshop Hawaii International Conference on System Science* (HICSS), Waikoloa, HI.

Scholtz, J., Young, J.L., Drury, J., and Yanco, H.A. (2004). Evaluation of human-robot interaction awareness in search and rescue. Paper presented at the *International Conference on Robotics and Automation (ICRA).* April 26-May 1, New Orleans, LA.

Schott, C. (2011). *A National Day for EOD.* U.S. Air Force. Available at: http://www.barksdale.af.mil/.

Schultze, U. and Leahy, M.M. (2009). The avatar self relationship: Enacting presence in Second Life. International Conference on Information Systems

(ICIS). *ICIS 2009 Proceedings*, Paper 12. Available at: http://aisel.aisnet.org/icis2009/12.

Schunk, D.H. (2000). *Learning Theories: An Educational Perspective.* Upper Saddle River, NJ: Prentice Hall.

Scott, R. (Director). (1982). *Blade Runner* (Five-Disc Complete Collector's Edition). [Motion picture]. United States: Warner Brothers.

Sesana, L. (2013, January 10). *Military Working Dogs have Long History of Heroism.* The Washington Times Communities. Available at: http://communities.washingtontimes.com/neighborhood/world-our-backyard/2013/jan/11/military-working-dogs-today/.

Shaker, S.M. (2011, July 11). Robot race to the moon. *COSMOS Magazine Online.* Available at: http://www.cosmosmagazine.com.

Shelley, M.W. (1998). *Frankenstein: Or, the Modern Prometheus: The 1818 Text* (World's Classics). Oxford: Oxford University Press.

Siebold, G. (2007). The essence of military group cohesion. *Peace Research Abstracts Journal*, 44(2), 286.

Silverstein, J. (2010, December 20). *Able to Rescue Wounded Soldiers and Aid the Elderly and Infirm, Robots Come of Age.* ABC News. Available at: http://abcnews.go.com/Technology/story?id=2740699&page=1.

Silverstein, S. (1996). My robot. In *Falling up.* New York: Harper Collins.

Singer, P.W. (2009). *Wired for War: The Robotics Revolution and Conflict in the 21st Century.* New York: Penguin Books.

Singer, P.W. (2010). How the U.S. Military can win the robot revolution. *Popular Mechanics.* Available at: http://www.popularmechanics.com/technology/robots/how-to-win-military-revolution.

Slagle, M. (2007, September 12). *Company Present a Boyish Robot.* The Washington Post. Available at: http://www.washingtonpost.com/wp-dyn/content/article/2007/09/12/AR2007091201969_pf.html.

Spencer, E.H. (2011, July-August). Raising Army EOD entry requirements. *Army Sustainment*, 43(4), PB-700-11-04. Available at: http://www.almc.army.mil/alog/issues/JulAug11/.

Staal, M.A. (2004). *Stress, Cognition, and Human Performance: A Literature Review and Conceptual Framework.* NASA/TM–2004–212824. Moffett Field, CA: NASA/Ames Research Center.

Stebbins, R.A. (2001). *Exploratory research in the social sciences.* Thousand Oaks, CA: Sage.

Stefanovich, J. (2002, July-August). Operation Enduring Freedom: EOD Operations. *News from the Front.* Ft. Leavenworth, KS: Center for Army Lessons Learned.

Steiner, I.D. (1972). *Group Process and Productivity.* New York: Academic Press.

Stewart, G.L., and Barrick, M.R. (2000). Team structure and performance: Assessing the mediating role of intrateam process and the moderating role of task type. *Academy of Management Journal*, 43(2), 135–48.

Stokes, A.F. and Kite, K. (2001). On grasping a nettle and becoming emotional. In P.A. Hancock, and P.A. Desmond (eds), *Stress, Workload, and Fatigue*. Mahwah, NJ: L. Erlbaum.

Straube, T., Preissler, S., Lipka, J., Hewig, J., Mentzel, H.J., and Miltner, W.H. (2010). Neural representation of anxiety and personality during exposure to anxiety-provoking and neutral scenes from scary movies. *Human Brain Mapping*, 31(1), 36–47.

Strauss, A. and Corbin, J. (1990). *Basics of Qualitative Research: Grounded Theory Procedures and Techniques*. Newbury Park, CA: Sage.

Strong, J.S. (1994). *The Legend and Cult of Upagupta: Sanskrit Buddhism in North India and Southeast Asia*. New Delhi: Motilal Banarsidass.

Suciu, P. (2013, April 8). Pentagon's Terminator-like robot to test military gear. *RedOrbit*. Available at: http://www.redorbit.com/news/technology/1112818232/humanoid-robot-boston-dynamics-040813/.

Sundstrom, E. and Altman, I. (1989). Physical environments and work group effectiveness. In L. Cummings and B. Staw (eds), *Research in Organizational Behavior*, 11, 175–209. Philadelphia, PA: Elsevier.

Sundstrom, E., De Meuse, K.P., and Futrell, D. (1990). Work teams: Applications and effectiveness. *American Psychologist*, 45(2), 120–33. doi:10.1037/0003-066X.45.2.120.

Sung, J.Y., Guo, L., Grinter, R.E., and Christensen, H.I. (2007). "My Roomba is Rambo": Intimate home appliances. In J. Krumm et al., (eds) *Proceedings of Ubicomp 2007*, LNCS 4717, 145–62. Berlin: Springer.

Svan, J. (2008, June 1). Bomb disposal: Same job, different pay. *Stars and Stripes*. Available at: http://www.stripes.com/news/.

Talton, T. (2008, June 20). Corps seeks to end EOD technician shortage. *Marine Corps Times*. Available at: http://www.marinecorpstimes.com/news/2008/06.

Trumbull, D. (Director) (1972). *Silent Running* [Film]. In M. Gruskoff (Producer). USA: Universal Studios.

UN News Centre. (2013). *UN Human Rights Expert Urges Global Pause in Creation of Robots with 'Power to Kill'* [Press release]. Available at: http://www.un.org/apps/news/story.asp?NewsID=45042.

UNAMA. (2012). *UNAMA Condemns Civilian Casualties Caused by Illegal Pressure Plate IED and Urges Anti-government Elements to Cease their Use*. [Press release]. Available at: http://unama.unmissions.org.

U.S. Army (n.d.). *Soldier Life: Basic Combat Training*. Available at: http://www.goarmy.com/soldier-life/becoming-a-soldier/basic-combat-training.html.

U.S. Army (n.d.). *Careers and Jobs: Unmanned Aerial Vehicle Operator (15W)*. Available at: http://www.goarmy.com/careers-and-jobs/browse-career-and-job-categories/transportation-and-aviation/unmanned-aerial-vehicle-operator.html.

U.S. Army (1997). *Ordnance Corps Vision: America's Army of the 21st century*. Aberdeen Proving Ground, MD: U.S. Army Ordnance Center and School.

U.S. Army (2001, June 14). *Department of the Army Publication FM 3-0.* Washington, D.C.

U.S. Army (Producer) (2010). *Inside look at 89 D–Explosive Ordinance Disposal (EOD) Specialist.* [Recruitment video]. United States: U.S. Army. Available at: https://youtu.be/hNQP4UU0EKo.

Vowell, M. (2013, March 28). EOD tech excels in stressful job. *Fort Campbell Courier.* Available at: http://www.fortcampbellcourier.com/news/.

Walker, L. (2015, March 8). Japan's robot dogs get funerals as Sony looks away. *Newsweek.* Retrived from http://www.newsweek.com/japans-robot-dogs-get-funerals-sony-looks-away-312192.

Webster, A. (2012, April 3). Navy robot training center simulates life in the desert and jungle. *TheVerge.* Retrieved http://www.theverge.com/2012/4/3/2922157/nrl-navy-robot-training-center.

Wilson, C.C. (1994). Commentary: A conceptual framework for human-animal interaction research: The challenge revisited. *Anthrozoös,* 7(1), 4–24.

Wilson, C. (2007, November 21). *Improvised Explosive Devices (IEDs) in Iraq and Afghanistan: Effects and Countermeasures.* CRS Report for Congress.

Wong, L., Kolditz, T., Millen, R., and Potter, T. (2003). *Why they Fight: Combat motivation in the Iraq War.* Carlisle, PA: Strategic Studies Institute.

Woods, D.D., Tittle, J., Feil, M., and Roesler, A. (2004, May 1). Envisioning human-robot coordination in future operations. *Proceedings of IEEE Transactions on Systems, Man and Cybernetics: Part C--Applications and Reviews,* 34(2), 210–18. doi:10.1109/TSMCC.2004.826272.

Valentin, E.A. (2011, February 3). Iraqi police improve IED skills. *On Patrol: The Magazine of the USO.* Available at: http://usoonpatrol.org/archives/2011/02/03/.

Veruggio, G. (2006, December 4–6). Euron Robotethics roadmap. *2006 6th IEEE-RAS Conference on Humanoid Robots,* 2, 612–17. doi:10.1109/ICHR.2006.321337.

Viskovatoff, A. (1999). Foundations of Niklas Luhmann's Theory of Social Systems. *Philosophy of the Social Sciences,* 29(4). doi: 10.1177/004839319902900402.

von Bertalanffy, L. (1968). *General Systems Theory.* New York: G. Braziller.

Vygotsky, L.S. (1986). *Thought and Language.* In A Kozulin (Trans. and Ed.) Ed. Cambridge, MA: MIT Press.

Yanco, H.A. and Drury, J. (2004). Classifying human-robot interaction: an updated taxonomy. *Proceedings of Systems, Man and Cybernetics, 2004 IEEE International Conference,* 3, 2841–6. New York: IEEE.

Yarbrough, B. (2008, February 1). Brotherhood of the bomb, Explosive Ordnance Disposal units: The Army's emergency responders. *Hesperia Star.* Retrieved http://www.hesperiastar.com/news/eod-1521-afghanistan-simeroth.html.

Yoon-Mi, K. (2007, April 29). Korea drafts robot ethics charter. *The Korea Herald.* Available at: www.koreaherald.co.kr.

Yost, P. (1989, October 10). "Manny" walks, talks—but robot can't run: Army mechanical man has key role in chemical warfare test program. *The Washington Post*. Available at: http://articles.latimes.com/1989-10-08.

Index

Notes: italics indicate images; 'n' indicates footnotes.